**Lecture Notes in Statistics** **163**

Edited by P. Bickel, P. Diggle, S. Fienberg, K. Krickeberg, I. Olkin, N. Wermuth, and S. Zeger

Springer Science+Business Media, LLC

Erkki P. Liski
Nripes K. Mandal
Kirti R. Shah
Bikas K. Sinha

# Topics in Optimal Design

Springer

Erkki P. Liski
Department of Mathematics, Statistics,
and Philosophy
University of Tampere
Kehruukoulunkatu 1
Tampere 33014
Finland

Nripes K. Mandal
Department of Statistics
Calcutta University
35, B.C. Road
Calcutta 700 019
India

Kirti R. Shah
Department of Statistics
and Actuarial Science
University of Waterloo
Waterloo, Ontario N2L 3G1
Canada

Bikas K. Sinha
Stat-Math Divison
Indian Statistical Institute
203 Barrackpore Trunk Road
Calcutta 700 035
India

Library of Congress Cataloging-in-Publication Data
Topics in optimal design / Erkki P. Liski ... [et al.].
p. cm. — (Lecture notes in statistics)
Includes bibliographical references and index.
ISBN 978-0-387-95348-9 ISBN 978-1-4613-0049-6 (eBook)
DOI 10.1007/978-1-4613-0049-6
1. Experimental design. I. Liski, Erkki. II. Lecture notes in statistics (Springer-Verlag); v. 163.
QA279 .T67 2001
519.5′3—dc21 2001049266

Printed on acid-free paper.



Camera ready copy provided by the authors.

9 8 7 6 5 4 3 2 1

ISBN 978-0-387-95348-9 SPIN 10849707

# Preface

In the early nineties, at the initiative of Sinha and financial support of Shah and Liski (from their respective Research Project Funds), the authors – inspired by their similar research interests – started collaborative research at various institutions mostly in pairs and triplets. It took more time and efforts on the part of Mandal to visit the others at regular intervals and keep track of their common as well as diverse research areas and merge his own. From this collaborative work, the concept of this monograph took a preliminary shape only last year and serious efforts were started to combine diverse avenues into one.

Admittedly, it took more time than expected to converge to a common platform regarding the contents and broad coverage of the topics to be included. We were mostly guided by our own common research interests spanning over the last ten years. That covered optimal designs in both discrete and continuous settings.

Availability of huge published literature in various statistical journals on the broad theme of optimal designs has made our task quite interesting and stimulating. We hope our readers will be as excited and delighted to read the monograph as we have been in our efforts to write it.

## Acknowledgments

We would like to thank Professor A. C. Mukhopadhyay (Indian Statistical Institute, Calcutta), whose critical comments on the first few chapters of an earlier draft have benefited us very much. Also it has been our pleasure to have Dr. Arto Luoma (University of Tampere, Finland) as our youngest collaborator in some research projects. His computing skills have surpassed all our expectations. We would also like to record our appreciation for the efforts given by Mrs. Helmi Untinen, University of Tampere, for the Latex-typing of certain parts of various versions of this Monograph.

Finally we wish to acknowledge the support of the Natural Sciences and Engineering Research Council of Canada in the form of a Research Grant to

Shah, the financial support of the Academy of Finland and the University of Tampere. The research visits of Mandal, Shah and Sinha were supported by the Academy of Finland Project no. 38113 and the University of Tampere. We are also grateful to Dr. Tapio Nummi and wish to acknowledge the support through his project (the Academy of Finland project no. 173691) for financing the latest visit of Sinha.

June 2001

Erkki P. Liski
Nripes K. Mandal
Kirti R. Shah
Bikas K. Sinha

# Personal Acknowledgments

To my co-authors Kirti R. Shah and Bikas K. Sinha for their hospitality when I visited Waterloo and Chicago. To my wife Leena and children Antti, Anni and Eero for promoting a pleasant environment towards my devotion to research.

Erkki P. Liski

To Professor S. K. Chatterjee, my revered teacher, under whose guidance I started working on optimum designs. To Professor Bikas K. Sinha, my respected teacher and co-author, who inspired me at every stage during the preparation of this monograph. To Professors Erkki P. Liski and Kirti R. Shah, my co-authors, for their hospitality and co-operation during my visits in Finland and Canada.

To my wife Sadhana and children Madhura and Pradip for allowing me to work without demanding much of my time and attention during the preparation of this monograph.

Nripes K. Mandal

To my wife Daksha for providing a pleasant working environment and for entertaining my colleagues in crime when they visited Waterloo.

To the Department of Statistics and Actuarial Science at the University of Waterloo for all the facilities provided. To my co-authors for all their hospitality when I visited them.

Kirti R. Shah

To my co-authors Erkki P. Liski and Kirti R. Shah and their families - for their excellent hospitality at Tampere, Finland and at Waterloo, Canada during the many seasons I have spent away from home. To my third co-author Nripes K. Mandal – for bearing with my temperaments with extreme courtesy at all stages!

To my children Karabi and Kuver - who were constantly claiming to have grown up enough to support their mother unhesitatingly in all matters, and to my wife Pritha – who took all the trouble during these years to raise two wonderful kids with exemplary courage and patience. Hence – to my family for always having been supportive of me, my work and my passion.

Bikas K. Sinha

# Contents

# 1

# Scope of the Monograph

## 1.1 Introduction and Literature Review

This monograph covers recent research findings in certain areas of discrete and continuous optimal designs, largely touching upon the interests of the authors spanning over the past ten years or so. We hope that the reader will find this useful for pursuing further research on these topics.

The earliest references to the classical theory of regression designs and related optimality results in the literature can be found e.g. in Smith (1918) and Plackett and Burman (1946). Subsequent work was done by Elfving (1952,1955,1956,1959), Chernoff (1953), Lindley (1956) and others. Elfving's paper (1952) on *Optimum allocation in linear regression theory* can be regarded as the beginning of the optimality theory of regression designs. During the past thirty years the theory of optimal designs - both discrete and continuous - has developed rapidly and this has resulted in publication of several excellent books. On continuous designs we have books by Fedorov (1972), Silvey (1980), Pázman (1986), Atkinson and Donev (1992), and the latest by Pukelsheim (1993) and Schwabe (1996). Discrete optimal designs are covered in the monograph by Shah and Sinha (1989).

The concept of continuous (i.e. approximate) design was first introduced by Kiefer and Wolfowitz (1959). They studied D-optimality in a systematic manner. Subsequently, Kiefer and Wolfowitz (1960) obtained the celebrated *Equivalence theorem* which connects the D-optimality criterion with the G-optimality criterion. This theorem incidentally provides a tool for verifying whether a given design is D-optimal or not. This tool has been extensively used by others, subsequently leading to a vast development of the subject matter. Later, Kiefer (1961, 1962) extended the notion of D-optimality to $D_s$-optimality criterion when one is interested in a subset of parameters, and established the corresponding equivalence theorem. Silvey (1978) and Pukelsheim (1980) studied the problem of singularity in this situation. Fedorov (1972) extended the notion of A-optimality to linear optimality criterion and derived the corresponding equivalence theorem.

Kiefer (1974) unified the different optimality criteria to a single optimality criterion in terms of a function which satisfies some general properties. He termed such a function as $\phi$-optimality criterion. He also established equivalence theorem for $\phi$-optimality. This contribution has great impact because often in practice the criterion function is not of the classical type. Subsequently, contributions of numerous researchers have enriched this field of research.

Applications of optimality criteria to various design problems have been made by a number of authors. Hoel (1958,1965) used the D-optimality criterion in the context of estimating all parameters in polynomial regression under a single factor and a specified type of two-factor model, respectively. Guest (1958) used the minimax criterion for obtaining optimal design for polynomial regression. Karlin and Studden (1966) also gave a solution for D-optimal design in the case of polynomial regression under different weight functions. Some applications can also be found in Atkinson and Donev (1992), and in Pázman (1986).

Pukelsheim (1993) gave a comprehensive account of the available results in the context of the continuous theory. One school of researchers developed algorithms for construction of optimal regression designs; notable contributions have been made by Wynn (1972), Fedorov (1972), Gaffke and Heiligers (1996) and others. A useful reference is the Handbook of Statistics, Vol. 13 (Ghosh and Rao 1996, eds.). Another development is associated with the name of G. E. P. Box who developed methods primarily for tackling applied problems. His work is reported, for example, in papers and books by Box and Wilson (1951), Box and Draper (1959,1987) and Box (1985).

Yet another school of researchers, primarily being motivated by Bayesian inference, has initiated studies on optimal designs based on the analysis of posteriors. The set-up has a built-in prior knowledge of some or all of the parameters in the form of prior diatribution(s) and the purpose is to optimize suitable functionals of the posterior information matrix of the parameters of interest. This area of research is also quite fascinating and at times useful from practical point of view. We will *not* discuss any aspect of what are known as Bayesian optimal designs. An excellent and updated review is to be found in Chaloner and Verdinelli (1995).

In finding an optimal design one determines first an *optimal information matrix* (in some sense) and then characterizes a design with the desired information matrix. Towards attaining such specified information matrices, Caratheodory's theorem gives a bound on the number of support points of the underlying designs. It says that for every design $d$ there exists a design $\tilde{d}$ which has the same information matrix as $d$ but $\tilde{d}$ has at most $m(m+1)/2+1$ support points where $m$ is the order of the information matrix, i.e. in the linear model $(\mathbf{Y}, \mathbf{X}\boldsymbol{\beta}, \sigma^2\mathbf{I})$ the parameter vector $\boldsymbol{\beta}$ is of order $m$.

In the special case of polynomial regression, a result to be referred to as *de la Garza phenomenon* (de la Garza 1954), abbreviated as DLG phenomenon, gives a much more powerful result. It says that for every single-factor polynomial regression design $d$ in a homoscedastic set-up, there exists a $(k+1)$-point design $\tilde{d}$ such that $d$ and $\tilde{d}$ have identical information matrices where $k$ is the degree of the polynomial. In this monograph we re-visit the DLG phenomenon in Chapters 2 and 3, making repeated use of it.

It must be noted, however, that the optimality results on polynomial regression designs available in the literature deal mostly with symmetric experimental domains. We refer to Pukelsheim (1993). Notable exceptions are certain examples in Pázman (1986) and an optimality result reported in Atkinson and Donev (1992). Symmetry plays a crucial simplifying role in the characterization of optimal designs. With a symmetric experimental domain, one can exploit considerations of symmetry, invariance and convexity/concavity to simplify the structure of competing designs, thereby reducing the dimension of the problem in the search of an optimal design. In Chapter 2, we present some results in this direction. Without appropriate symmetry of the experimental domain, however, much of what has been studied does *not* seem to have an obvious generalization.

In Chapter 3 of this monograph we deal with the problem of characterization of optimal designs in the set-up of linear, quadratic and cubic regression for asymmetric experimental domains. Here the results for symmetric experimental domains are inapplicable, and therefore, we do not obtain much simplification in the optimality considerations. With this background in mind we have exploited the DLG phenomenon and Loewner order domination (LOD) of information matrices in the search of an optimal design. This leads to simplification of a very general nature and specific optimal designs can be studied with ease. We deal with the fixed and/or random coefficient linear, quadratic and cubic regression models. In the other chapters of this monograph we deal with both discrete and continuous optimal designs for specific model-related problems of linear inference. In Section 1.4 we present a summary of the monograph.

## 1.2 Some Useful Linear Models

In this section we present a short overview of linear *fixed-* and *mixed-effects models* which have an important role in this monograph. Many important statistical models can be expressed as mixed effects models that incorporate both *fixed- effects* and *random-effects.* Examples of the underlying datasets include *longitudinal data, repeated measures data, multilevel data* and *block designs.*

At first we review a polynomial regression model with fixed regression coefficients and introduce the concept of an approximate (i.e., continuous) design in the process. Next we formulate a general *mixed linear model,* and then two important special cases of it: a *random coefficient regression* (*RCR*) model and a *growth curve (GC)* model. The theory of optimal designs so far primarily deals with models where observations are independent and identically distributed. In this monograph we derive various results under a more general setting. Especially important is the optimality theory under certain RCR models. The classical results of fixed effects regression follow as a special case from the theory for RCR models. Although we present substantial genalizations of the classical linear model design theory, a general optimal design theory for mixed linear models is still missing. There are plenty of open problems in this field of research.

### 1.2.1 A Polynomial Regression Model: Exact and Approximate Theory

We consider a $k$th degree polynomial regression model where we have a set of uncorrelated responses

$$Y_{ij} = \beta_0 + \beta_1 x_i + \beta_2 x_i^2 + \ldots + \beta_k x_i^k + e_{ij} \tag{1.2.1}$$

for $i = 1, 2, \ldots, n$ and $j = 1, 2, \ldots, N_i$ with expectations and variances given by

$$E(Y_{ij}) = \beta_0 + \beta_1 x_i + \beta_2 x_i^2 + \ldots + \beta_k x_i^k, \quad V(Y_{ij}) = \sigma^2 \tag{1.2.2}$$

respectively. Suppose for simplicity that $\sigma^2 = 1$. The *dispersion matrix*

$$\mathbf{V}(\tilde{\beta}) = \frac{1}{N}\left(\sum \frac{N_i}{N} x_i^{r+s-2}\right)^{-1}_{1 \le r,s \le k+1} \tag{1.2.3}$$

of the least squares estimate (LSE) of $\beta = (\beta_0, \beta_1, \beta_2, \cdots, \beta_k)'$ depends on the values $x_1, x_2, \ldots, x_n$ of $x$ and also on the number of replications $N_1, N_2, \ldots, N_n$, given that $N = N_1 + N_2 + \ldots + N_n$ refers to the total number of observations.

The *experimental conditions* $\{x_1, x_2, \ldots, x_n\}$ are assumed to lie in the interval $\mathcal{T} = [a, b]$, which is called the *experimental domain.* (We will deal with various forms of $\mathcal{T}$ in this monograph - with special reference to *symmetric* and *asymmetric* experimental domains.) The corresponding *regression range*

$$\chi = [(1, x, x^2, \cdots, x^k)' : x \in \mathcal{T}]$$

is a $(k+1)$-dimensional subset of vectors in $\mathbf{R}^{k+1}$. An *experimental design*

$$d_{n:N} = \{x_1, x_2, \ldots, x_n; N_1, N_2, \ldots, N_n\} \tag{1.2.4}$$

or simply $d_n$, for a sample of size $N$ is given by a finite number $n$ ($\geq k+1$) of *distinct* regression values in $\mathcal{T}$, and positive integers $N_1, N_2, \ldots, N_n$ with $N_1 + N_2 + \ldots + N_n = N$. The set of points $x_1, x_2, \ldots, x_n$ or the corresponding *regression vectors* $\mathbf{x}_i = (1, x_i, x_i^2, \ldots, x_i^k)'$, $i = 1, 2, \ldots, n$ are called the *support* of $d_n$. Designs of the type $d_{n:N}$ for a specified number of trials $N$ are called exact and the allocation numbers are specified as in (1.2.4).

Any collection

$$d_n = \{x_1, x_2, \ldots, x_n; p_1, p_2, \ldots, p_n\} \tag{1.2.5}$$

of $n$ ($\geq k+1$) distinct points $x_i \in \mathcal{T}$ and positive numbers $p_i$, $i = 1, 2, \ldots, n$ such that $\sum p_i = 1$, induces an *approximate design* $d$ on the experimental domain $\mathcal{T}$. Sometimes it is also referred to as a continuous design (vide Pukelsheim 1993, p. 32). For a given *continuous design* $d_n$ of the form in (1.2.5), the *information matrix* (on a per observation basis) for the parameters in (1.2.1) is defined by

$$\mathcal{I}(\boldsymbol{\beta}) = \sum p_i \mathbf{x}_i \mathbf{x}_i' = (\mu_{r+s-2})_{1 \leq r,s \leq k+1} \tag{1.2.6}$$

where

$$\mu_{r+s-2} = \sum p_i x_i^{r+s-2}. \tag{1.2.7}$$

This per unit information matrix of an *exact design* is $\frac{1}{N}\mathbf{V}^{-1}(\tilde{\boldsymbol{\beta}})$ so that $\mathbf{V}(\tilde{\boldsymbol{\beta}})$ can be expressed as $\frac{1}{N}\mathcal{I}^{-1}(\boldsymbol{\beta})$ with the interpretation that $p_i = N_i/N$, $i = 1, 2, \ldots, n$. Naturally, for an exact design, the weights $p_i = N_i/N$ can have only a finite number of distinct values in $(0,1)$. In practice, all designs are exact in the sense of (1.2.4). However, for large values of $N$, $d_n$ in (1.2.5) may be regarded as an approximation to (1.2.4). In this monograph, we will discuss only continuous designs while dealing with the set-up of polynomial regression models.

Before concluding this section, we will briefly mention two other points. These refer to (*i*) a subset of parameters and (*ii*) *de la Garza phenomenon* respectively. If we write $\boldsymbol{\beta} = (\boldsymbol{\beta}^{(1)'}, \boldsymbol{\beta}^{(2)'})'$, then for a subset of the parameters, say for $\boldsymbol{\beta}^{(2)}$ the information matrix is given by

$$\begin{aligned} \mathcal{I}(\boldsymbol{\beta}^{(2)}) &= \mathcal{I}_{22.11} \\ &= \mathcal{I}_{22} - \mathcal{I}_{21}\mathcal{I}_{11}^{-}\mathcal{I}_{12} \end{aligned} \tag{1.2.8}$$

where $\mathcal{I}$ has the natural partitioning:

$$\mathcal{I} = \begin{pmatrix} \mathcal{I}_{11} & \mathcal{I}_{12} \\ \mathcal{I}_{21} & \mathcal{I}_{22} \end{pmatrix}. \tag{1.2.9}$$

Next consider a set of *linear combinations of parameters*, say $\boldsymbol{\eta} = \mathbf{L}\boldsymbol{\beta}$ with full row rank of $\mathbf{L}$. In order to find the information matrix for $\boldsymbol{\eta}$, we can introduce $\boldsymbol{\xi} = \mathbf{M}\boldsymbol{\beta}$ so that $\mathbf{M}$ is also of full row rank and further, $\{\mathbf{L}', \mathbf{M}'\}$ is square non-singular. That way, $\boldsymbol{\beta}^* = (\boldsymbol{\xi}', \boldsymbol{\eta}')'$ defines a non-singular transform of $\boldsymbol{\beta}$. We can now define $\mathcal{I}^*(\boldsymbol{\beta}^*)$ by changing the model parameters from $\boldsymbol{\beta}$ to $\boldsymbol{\beta}^*$. Finally, we will work out $\mathcal{I}(\boldsymbol{\eta})$ as $\mathcal{I}^*_{22.11}$.

In the set-up of polynomial regression model (1.2.1), the design $d_n$ in (1.2.5) and the information matrix $\mathcal{I}(\boldsymbol{\beta})$ in (1.2.6), we now describe the de la Garza (DLG) phenomenon. This deals with *information equivalence*. It states that for the model (1.2.1), given $d_n$ in (1.2.5), there exists a design $d^*_{k+1}$ of the type

$$d^*_{k+1} = \{x^*_1, x^*_2, \ldots, x^*_{k+1};\ p^*_1, p^*_2, \ldots, p^*_{k+1}\} \tag{1.2.10}$$

which is based on precisely $k+1$ *support points* such that $\mathcal{I}(d_n) = \mathcal{I}(d^*_{k+1})$ where $\mathcal{I}(d)$ stands for $\mathcal{I}(\boldsymbol{\beta})$ evaluated for the design $d$. Moreover, $x_{\min} \leq$

$x^*_{\min} < x^*_{\max} \leq x_{\max}$ (de la Garza 1954). We will *not* make any attempt to provide a proof of the DLG Phenomenon in its most general form. However, the cases of linear and quadratic regression will be taken up in Chapter 3. Also, in Chapter 2, it will be discussed for certain classes of designs under a quadratic regression model.

In Chapters 2 and 3, we will mostly exploit the DLG phenomenon to characterize and construct optimal regression designs when the experimental domain may be asymmetric in nature.

### 1.2.2 A General Mixed Linear Model

Let $K$ denote the number of *experimental units* and $\mathbf{y}_i$ be the $n_i \times 1$ vector of measurements on the $i$th unit. We assume the mixed model

$$\mathbf{y}_i = \mathbf{X}_i\boldsymbol{\beta} + \mathbf{Z}_i\mathbf{b}_i + \mathbf{e}_i, \quad i = 1, 2, \ldots, K \tag{1.2.11}$$

where $\mathbf{y}_i$ is independent of $\mathbf{y}_j$ for all $i \neq j$, $\mathbf{X}_i$ and $\mathbf{Z}_i$ are known $n_i \times p$ and $n_i \times m$ design matrices and $\boldsymbol{\beta}$ is a vector of $p$ fixed *regression parameters*, $\mathbf{b}_i$ and $\mathbf{e}_i$ are independent $m \times 1$ and $n_i \times 1$ random vectors distributed as $N(\mathbf{0}, \mathbf{D})$ and $N(\mathbf{0}, \mathbf{R}_i)$, respectively. It follows that, marginally, $\mathbf{y}_i$'s are independent *normally distributed* random vectors with

$$E(\mathbf{y}_i) = \mathbf{X}_i\boldsymbol{\beta} \tag{1.2.12}$$

and

$$\mathbf{V}(\mathbf{y}_i) = \boldsymbol{\Sigma}_i = \mathbf{Z}_i\mathbf{D}\mathbf{Z}_i' + \mathbf{R}_i \tag{1.2.13}$$

where $\mathbf{D}$ and $\mathbf{R}_i$ are known/unknown *positive definite* $m \times m$ and $n_i \times n_i$ matrices, respectively, and $\mathbf{R}_i$ does not depend on $i$ except for its size i.e., it has the same structure for each $i$. The mixed model for measures was introduced by Laird and Ware (1982) in the case where $\mathbf{R}_i = \sigma^2\mathbf{I}_i$, $\mathbf{I}_i$ being the $n_i \times n_i$ identity matrix. Subsequently Lindstrom and Bates (1988) considered the more general model (1.2.13). The literature of mixed models is extensive and the basic results can be found in Searle (1971, 1987) and Searle *et al.* (1992).

In many situations data arise from a *randomized experimental* design where several successive measurements over time or space are made on each experimental unit. Modelling the behaviour *of repeated measurements*, also called *longitudinal data*, has been the subject of lively discussion especially in the biometrical literature. The books by Diggle (1994) and by Lindsay (1993) give a good overview of modelling and analysis of repeated measurements and longitudinal data. In these books there is a strong emphasis on applications in biological and health sciences.

In econometric literature (vide Baltagi 1995) the term "panel data" refers to the pooling of observations on a *cross-section* of households, countries, firms etc. over several time periods. Most of the panel data can be analyzed utilizing mixed effects models which combine fixed and random effects. In a series of papers Liski and Nummi (1995a and 1995b, 1996a and 1996b) have used a mixed model approach in various engineering and forestry applications. General algorithms for estimating the parameters in mixed effects models for repeated-measurements data are discussed by Jennrich and Schluchter (1986), Laird *et al.* (1987), Lindstrom and Bates (1988), Jones and Ackerson (1990) and Jones and Boadi-Boateng (1991).

We will now specialize to a random coefficient regression model (RCR) Model.

### 1.2.3 A Random Coefficient Regression (RCR) Model

The *random effects* $\mathbf{b}_1, \mathbf{b}_2, \ldots, \mathbf{b}_K$ in model (1.2.11) emphasize the variability of responses across units, instead of the homogeneity of responses on a unit. In random coefficient regression $\mathbf{Z}_i = \mathbf{X}_i$, or $\mathbf{Z}_i$ consists of a subset of the columns of $\mathbf{X}_i$ (vide Swamy 1971; Carter and Yang 1986). Perhaps the simplest example of this is when $\mathbf{Z}_i = \mathbf{1}_i$ is the $n_i \times 1$ vector of 1's, with $m = 1$. Then the general level of the response profile varies between units so that some units are intrinsically high responders, others low responders. This simple *RCR model* model is frequently encountered in practice.

Consider first, for simplicity, the *first-degree polynomial model.* Then for the $i$th unit at the point $x_{ij}$

$$Y_{ij} = (\beta_0 + b_{0i}) + (\beta_1 + b_{1i})x_{ij} + e_{ij} \tag{1.2.14}$$

with fixed mean line parameters $\beta_0$ and $\beta_1$ and with respective random effects $b_{0i}$ and $b_{1i}$ associated with the $i$th unit. Then under the mixed model formulation (1.2.11)

$$\mathbf{X}_i = \mathbf{Z}_i = \begin{pmatrix} 1 & 1 & \ldots & 1 \\ x_{i1} & x_{i2} & \ldots & x_{in_i} \end{pmatrix}' \tag{1.2.15}$$

$\mathbf{b}_i = (b_{0i}, b_{1i})'$ and $\boldsymbol{\beta} = (\beta_0, \beta_1)'$. The model (1.2.14) can easily be generalized to higher order polynomials.

The simplest *covariance structure*

$$\boldsymbol{\Sigma}_i = \sigma^2 \mathbf{I}_{n_i} \tag{1.2.16}$$

arises when there are no random effects. If we specify only one random effect $b_{0i}$, then $\mathbf{Z}_i = \mathbf{1}_i$ and we obtain the *uniform structure*

$$\boldsymbol{\Sigma}_i = \sigma_0^2 \mathbf{J}_i + \sigma^2 \mathbf{I}_{n_i}, \tag{1.2.17}$$

where $\sigma_0^2 = V(b_{i0})$ and $\mathbf{J}_i = \mathbf{1}_i \mathbf{1}_i'$, $\mathbf{1}_i$ being an $n_i \times 1$ vector of 1's. If $m = p$ and $\mathbf{Z}_i = \mathbf{X}_i$ in the RCR model, we have the saturated random effects model with

$$\boldsymbol{\Sigma}_i = \mathbf{X}_i \mathbf{D} \mathbf{X}_i' + \sigma^2 \mathbf{I}_{n_i}. \tag{1.2.18}$$

The model (1.2.14) is of this type.

Liski and Nummi (1995a, 1995b, 1996a and 1996b) have applied RCR in modelling of stem curves. It is well known in forestry that stand density, site type, climate and genetical factors affect the form of the stem curve. However, the mathematical models presented for stem curves do not usually contain these factors, since in practice they are difficult or impossible to measure. An apparent feature of timber is that stem curves typically vary randomly from tree to tree. This variation can be explained, at least partially, by the factors mentioned above. Although these factors are not explicitly included in the model, the RCR model accommodates this variation rather well. Polynomial models are simple and describe stem curves adequately. Even second- and third-degree polynomials fit the stem data fairly precisely. Liski *et al.* (1996, 1998) presented optimal designs for slope parameter estimation and prediction in a first degree RCR model. The results were applied to tree stem data collected by forest harvesters.

### 1.2.4 A Growth Curve Formulation

The *linear growth curve model* for serial measurements has a central role in longitudinal studies. The mixed model can also be seen to have origins in the growth curve literature, where the approach to modelling has been slightly different. The growth curve model was introduced by Potthoff and Roy (1964). The random coefficient growth curve model was originally proposed by Elston and Grizzle (1962). The growth curve model with random effects was developed by Rao (vide, Rao 1959, 1965 and 1967). Khatri (1966) and Grizzle and Allen (1969) extended Rao's work.

For the growth curve formulation, we assume for each individual unit $i$ the model

$$\mathbf{y}_i = \mathbf{Z}_i\mathbf{a}_i + \mathbf{e}_i, \quad i = 1, 2, \ldots, K \tag{1.2.19}$$

where $\mathbf{y}_i$, $\mathbf{e}_i$ and $\mathbf{Z}_i$ are as in the model (1.2.11). The component $\mathbf{Z}_i\mathbf{a}_i$ defines the $i$th individual's growth curve. Each $\mathbf{a}_i$ is a random parameter, unique to the $i$th individual. The parameters $\mathbf{a}_1$, $\mathbf{a}_2$, ..., $\mathbf{a}_K$ are independently distributed as $N(\mathbf{A}_i\boldsymbol{\beta}, \mathbf{D})$, where $\boldsymbol{\beta}$ and $\mathbf{D}$ are as previously defined, and $\mathbf{A}_i$ is an $m \times p$ design matrix. This implies that

$$E(\mathbf{y}_i) = \mathbf{Z}_i\mathbf{A}_i\boldsymbol{\beta} = \mathbf{X}_i\boldsymbol{\beta} \tag{1.2.20}$$

and

$$\mathbf{V}(\mathbf{y}_i) = \mathbf{Z}_i\mathbf{D}\mathbf{Z}_i' + \mathbf{R}_i, \tag{1.2.21}$$

where $\mathbf{X}_i = \mathbf{Z}_i\mathbf{A}_i$. Now $\mathbf{b}_i = \mathbf{a}_i - \mathbf{A}_i\boldsymbol{\beta}$ is a deviation of an individual's growth curve parameters from the parameters of the population growth curve.

The generalized analysis of variance model (Potthoff and Roy 1964) arises as a special case when $\mathbf{Z}_i = \mathbf{Z}$ and $n_i = n$ for all $i$. The connection between the mixed model and the growth curve model of Potthoff and Roy is discussed in more detail in Laird *et al.* (1987), for example.

Abt *et al.* (1997) considered optimal designs in a linear growth curve model under the intraclass correlation model. They assumed, based on physical constraints, that the time points accessible for actual measurements of $y$ are given by

$$-k, -(k-1), -(k-2), \ldots, -2, -1, 0, 1, 2, \ldots, k-2, k-1, k.$$

Thus, the time points are assumed to be equally spaced which is usual in time series data. It was further assumed, based on cost considerations, that a total of $N$ observations can be taken and that no experimental unit can be *recalled* once it is released, so that it is actually made available (when called for) for one or more consecutive time points in the time scale. At any time point, as many individuals as needed can be observed, subject to the constraint discussed above.

This study (Abt *et al.* 1997) indicated that there is essentially no difference between the i.i.d. case and some spesific correlated structures of the errors. The two-point distribution with 50 % load at the extreme values turns out to be the best in almost all situations. In the second part of the above paper (Abt *et al.* 1998) the aspect of optimality was studied in the context of a *quadratic growth curve model.* Similar results were also found for the autocorrelation structure of the error term. These results will *not* be discussed in this monograph.

# 1.3 Estimation of Regression Parameters in RCR Models

We assume now in (1.2.11) that $\mathbf{X}_i = \mathbf{Z}_i = \mathbf{X}$ and $n_i = n$ for every $i = 1$, 2, ..., $K$. Then, we have the *RCR model*

$$\mathbf{y}_i = \mathbf{X}\boldsymbol{\beta} + \mathbf{X}\mathbf{b}_i + \mathbf{e}_i, \quad i = 1, 2, \ldots, K \tag{1.3.1}$$

where $\boldsymbol{\beta}$ is a fixed mean curve vector and $\mathbf{b}_1, \mathbf{b}_2, \ldots, \mathbf{b}_K$ are random effects. It follows that the *generalised least squares (GLS) estimator* of $\boldsymbol{\beta}$ is

$$\tilde{\boldsymbol{\beta}} = (\mathbf{X}'\boldsymbol{\Sigma}^{-1}\mathbf{X})^{-1}\mathbf{X}'\boldsymbol{\Sigma}^{-1}\bar{\mathbf{y}} \tag{1.3.2}$$

with $\bar{\mathbf{y}} = \frac{1}{K}\sum_{i=1}^{K}\mathbf{y}_i$, where

$$\boldsymbol{\Sigma} = \mathbf{X}\mathbf{D}\mathbf{X}' + \sigma^2\mathbf{I}_n \tag{1.3.3}$$

is the dispersion matrix $\mathbf{V}(\mathbf{y}_i)$ of $\mathbf{y}_i$ and $\mathbf{D} = \mathbf{V}(\mathbf{b}_i)$ is the dispersion matrix of $\mathbf{b}_i$ for all $i = 1, 2, \ldots, K$. Thus the dispersion matrix of $\tilde{\boldsymbol{\beta}}$ is

$$\mathbf{V}(\tilde{\boldsymbol{\beta}}) = \frac{\sigma^2}{K}\left(\mathbf{X}'\boldsymbol{\Sigma}^{-1}\mathbf{X}\right)^{-1} \tag{1.3.4}$$

and the corresponding information matrix is

$$\frac{K}{\sigma^2}(\mathbf{X}'\boldsymbol{\Sigma}^{-1}\mathbf{X}). \tag{1.3.5}$$

Note that the GLS estimator (1.3.2) is the *best linear unbiased (BLU) estimator* of $\boldsymbol{\beta}$.

The ordinary *fixed coefficient regression (FCR) model* corresponding to (1.3.1) is

$$\mathbf{y}_i = \mathbf{X}\boldsymbol{\beta} + \mathbf{e}_i, \quad i = 1, 2, \ldots, K \tag{1.3.6}$$

which follows from (1.3.1) by supposing that $V(\mathbf{b}_1) = V(\mathbf{b}_2) = \cdots = V(\mathbf{b}_K) = \mathbf{0}$. Since a design $d$ for estimation of $\boldsymbol{\beta}$ does not depend on $\sigma^2$ and $K$, we can put $\sigma^2 = 1$ and omit $K$ when comparing designs. Therefore we may denote in the sequel that

$$\boldsymbol{\Sigma} = \mathbf{X}\mathbf{D}\mathbf{X}' + \mathbf{I}_n \tag{1.3.7}$$

and

$$\boldsymbol{\mathcal{I}} = \mathbf{X}'\boldsymbol{\Sigma}^{-1}\mathbf{X} \tag{1.3.8}$$

is the information matrix of $\boldsymbol{\beta}$.

In fact, the model (1.3.1) is a special case of Rao's general mixed model (vide Rao 1967, p. 355). In this model the covariance matrix of observations is said to have "Rao's structure". Rao (1967) proved under his general mixed model that the GLS estimator [like (1.3.2)] and the ordinary least squares (OLS) estimator are identifical. It therefore follows immediately from Rao's result that the OLS estimator $\tilde{\boldsymbol{\beta}}_{\mathrm{OLS}} = (\mathbf{X}'\mathbf{X})^{-1}\mathbf{X}'\mathbf{y}$ and the GLS estimator (1.3.2) are identical. Since

$$\mathbf{V}(\tilde{\boldsymbol{\beta}}_{\mathrm{OLS}}) = (\mathbf{X}'\mathbf{X})^{-1}\mathbf{X}'\boldsymbol{\Sigma}\mathbf{X}(\mathbf{X}'\mathbf{X})^{-1} = \mathbf{D} + (\mathbf{X}'\mathbf{X})^{-1}, \tag{1.3.9}$$

then we have, of course, that

$$\mathbf{V}(\tilde{\boldsymbol{\beta}}) = \mathbf{D} + (\mathbf{X}'\mathbf{X})^{-1}. \tag{1.3.10}$$

We may note in passing that under the model (1.3.6), $\tilde{\boldsymbol{\beta}} = \tilde{\boldsymbol{\beta}}_{OLS}$ and $\mathbf{V}(\tilde{\boldsymbol{\beta}}) = (\mathbf{X}'\mathbf{X})^{-1}$. Puntanen and Styan (1989) presented a good overview, in a historical perspective, of the conditions under which the OLS estimator and the BLU estimator are identical.

We can now specialize to the case of RCR models in polynomial regression and work out the continuous design analogue of the information matrix in (1.2.6), taking the clue from (1.3.10). As before, we will consider a $k$th degree polynomial regression model but work with random coefficients for the regression parameters as in (1.2.14). The design $d_n$ in (1.2.5) represents a continuous design involving $n$ design points. Under a fixed effects model, the resulting information matrix is given by $\boldsymbol{\mathcal{I}}$ in (1.2.6) so that the dispersion matrix of the LSE's of the parameters is given by

$$\mathbf{V}(\tilde{\boldsymbol{\beta}}) = \boldsymbol{\mathcal{I}}^{-1} = (\mathbf{X}'\mathbf{D}_p\mathbf{X})^{-1} \tag{1.3.11}$$

where

$$\mathbf{X} = \begin{pmatrix} 1 & x_1 & \ldots & x_1^k \\ 1 & x_2 & \ldots & x_2^k \\ \vdots & \vdots & \ddots & \vdots \\ 1 & x_n & \ldots & x_n^k \end{pmatrix} \text{ and } \mathbf{D}_p = \mathrm{Diag}(p_1, p_2, \ldots, p_n). \tag{1.3.12}$$

From (1.3.10), it turns out that under RCR model, the dispersion matrix will change from (1.3.11) to

$$\mathbf{V}(\tilde{\beta}) = \mathbf{D} + (\mathbf{X}'\mathbf{D}_p\mathbf{X})^{-1}, \tag{1.3.13}$$

where $\mathbf{D}$ is a $(k+1)\times(k+1)$ diagonal matrix having the variance components of the random coefficients as its diagonal elements. In other words,

$$\mathbf{D} = \text{Diag}(\delta_0, \delta_1, \cdots, \delta_k),\ \delta_i = \mathbf{V}(b_i)/\sigma^2,\ i = 0, 1, \ldots, k. \tag{1.3.14}$$

Denoting by $\mathbf{V}_F$ and $\mathcal{I}_F$ the expressions for the dispersion matrix and the information matrix respectively resulting from a polynomial regression model with fixed regression parameters, and by $\mathbf{V}_R$ and $\mathcal{I}_R$ the corresponding expressions resulting from the RCR model, we now have the relations:

$$\mathbf{V}_R = \mathbf{D} + \mathbf{V}_F, \mathcal{I}_R = [\mathbf{D} + \mathcal{I}_F^{-1}]^{-1}. \tag{1.3.15}$$

The above representation clearly indicates that the DLG phenomenon holds trivially for RCR models in polynomial regression as well. Further, any information dominance in the fixed effects models will also carry through for random coefficients models. Various optimality aspects of linear and qudratic random coefficient regression models over an asymmetric experimental domain will be discussed in this monograph.

## 1.4 Chapter Summary

In Chapter 2 we describe Loewner ordering technique for comparing competing designs. We then give optimality results for quadratic regression where the set of possible values of the regressor variable is symmetric about the origin. We also give optimality results for multifactor linear regression model for the case where the experimental domain is either a Euclidean ball or hypercube.

In Chapter 3, we deal with characterization of specific optimal designs in linear and quadratic regression over $[0, 1]$ for estimation of parameters. Also we carry out the same for prediction and inverse prediction problems while the design range is $[0, h]$ and the prediction range is $[h, H]$. All this is done covering the situations where the regression coefficients are fixed or random. We also present the A-optimal designs over $[0, 1]$ under cubic regression with fixed regression coefficients.

In Chapter 4, we study the problem of efficient estimation of regression coefficients for all the covariates in a multiple linear regression model with the experimental domain being the product space of $[-1, 1]$ in each factor. However, the intercept term is supposed to be affected by possible different treatment effects and block effects. This can also be viewed as the problem of efficient estimation of treatment effects contrasts in a CRD model or a block design in the presence of a number of covariates, each ranging in $[-1, 1]$. The key reference to this Chapter is Lopes Troya

(1982). We investigate the above problem in the framework of a CRD, an RBD and a BIBD. Optimal designs depend heavily on the existence and use of Hadamard matrices and mutually orthogonal Latin squares (MOLS). The non-trivial problem is to accommodate as many covariates as possible without sacrificing information on any one of them or on the treatment effects.

In Chapter 5 we deal with optimality with respect to a criterion introduced in Sinha (1970). Since this is based on the *distance* between the parameter and its estimator in a *stochastic* sense, it is known as *distance optimality* or *DS-optimality* criterion. Okamoto's Lemma and its several generalizations provide useful tools in the study of DS-optimality. Several properties of this criterion are studied and its relation with D- and E-optimality criteria is noted.

Distance optimality is not an easily tractable one. However, it has been possible to obtain some tangible results. It is shown that CRD's with symmetrical or nearly symmetrical allocations are DS-optimal. BIBD's are also shown to be DS-optimal. In the regression set-up, uniform simplex designs are shown to be uniquely DS-optimal.

DS-optimal designs for comparing a test treatment with a set of control treatments are also studied in this context. CRDs which maximize coverage probabilities for a fixed distance ($\epsilon$) between the parameter vector and its estimator are obtained. Since the optimal designs depend upon the value of $\epsilon$, we also try to obtain designs which maximize the coverage probability weighted with respect to a probability distribution for $\epsilon$. The analytical results are supplemented by numerical computations which illustrate the nature of the optimal designs. Much of the work described in this chapter is in terms of approximate designs.

In Chapter 6 we study block designs where in each block there is a linear trend effect which may be different for different blocks. The problem was first studied by Bradley and Yeh (1980) who assumed a common trend and tried to obtain optimal trend-free designs i.e., designs for which the information matrix for treatment effects remains unaltered when these are adjusted for the trend parameter.

Initially, we present designs which are universally optimal (UO) within the restricted class of binary designs or are UO within the class of trend-free designs. Semi-balanced arrays (SBA's) introduced by Rao (1973) are found to be very helpful in the search for optimal designs in this setting. It may be remarked that all optimality results in this set-up are in terms of UO and this is accomplished by obtaining a completely symmetric information matrix with maximal trace - a well-known technique suggested by Kiefer (1975).

Next, we present designs which are UO in the unrestricted class. Main challenge here is in obtaining allocations which maximize the trace. In both cases i.e., for restricted class or for unrestricted class, the approach has been to start with a BIBD and to re-arrange the treatments in each block so that the trace of the information matrix (in the model with the presence of trend) is maximized. Finally, we give efficiency bounds in situations where exact optimal designs can not be obtained. As is usually the case, these relate to A-optimality. It may be noted that all work in this chapter relates to discrete (i.e. exact) optimal designs.

In Chapter 7, we deal with a number of topics. The approach here is to introduce the topic, give the flavor of the results available at this time and often to indicate directions for future research in the area. We start with a slightly modified framework for UO and indicate its relationship with Kiefer's framework. Next, we deal with the models with competing effects. Here again the results are in terms of UO.

In Section 7.3 we consider split block designs (vide Ozawa *et al.* 2001) and show that a balanced design (when it exists) is UO optimal for the estimation of the interaction effects. Next section deals with nested designs. Following Morgan (1996), a general framework is presented and bottom stratum analysis as well as full analysis are presented. We also comment on mixed models. Some known optimality results are presented. Much of the work presented is for nesting of one factor within the levels of another. We make a brief mention of the case where two nested factors are crossed within each level of a nesting factor.

We then deal with 3-way incomplete layouts. Work in this area was pioneered by Agrawal (1966) who gave various methods of construction for what are known as 3-way balanced designs i.e., designs providing completely symmetric **C**-matrices for each of the three classifications: rows, columns and treatments. In view of this extreme symmetry, it would be natural to expect that these designs are optimal. However, that is not the case. We present designs which are superior to Agrawal's designs for all the classifications! In some cases, E-optimal designs are also obtained. This section deals with optimal estimation of parametric contrasts in *each* of the classifications. All work here is in the framework where an *incomplete* row-column layout is given and the problem is of optimal allocation of treatments to the available experimental units.

In the last section we study a specific regression model where the errors are heteroscedastic. This arises in a Poisson count model studied, among others, by Minkin (1993).

In this monograph, whenever applicable, continuous design theory has been utilized for the derivation of optimal designs. We have *not* attempted presentation of any exact design analogoues of such optimal designs. As a thumb rule, "nearest integer" approximation has been the practice and it is likely to provide efficient designs. Pukelsheim (1993, Chapter 12) has studied this problem rigourously and presented some algorithms for tackling this aspect of nearest integer approximation. We will *not* discuss this issue any further in this monograph.

## References


Abt, M., Gaffke, N., Liski, E. P. and Sinha, Bikas K. (1998). Optimal designs in crowth curve models: Part II. Correlated model for quadratic growth: Optimal designs for slope parameter estimation and growth prediction. *Journal of Statistical Planning and Inference* **67**, 287–296.

Abt, M., Liski, E. P., Mandal N. K. and Sinha, Bikas K. (1997). Optimal design in crowth curve models: Part I. Correlated model for linear growth: Optimal designs for slope parameter estimation and growth prediction. *Journal of Statistical Planning and Inference* **64**, 141–150.

Agrawal, H. L. (1966). Some systematic methods of construction of designs for two-way elimination of heterogeneity. *Calcutta Statistical Association Bulletin* **15**, 93–108.

Atkinson, A. C. and Donev, A. N. (1992). *Optimum experimental design.* Oxford: Oxford University Press.

Baltagi, B. H. (1995). *Econometric analysis of panel data.* Wiley, New York.

Box, G. E. P. (1985). *The Collected works of George E. P. Box* (Eds. G. C. Tiao, C. W. J. Granger, I. Guttman, B. H. Margolin, R. D. Snee, S. M. Stigler). Wadsworth, Belmont, CA.

Box, G. E. P. and Draper, N. R. (1959). A basis for the selection of a response surface design. *Journal of the American Statistical Association* **54**, 622–654.

Box, G. E. P. and Draper, N. R. (1987). *Empirical model-building and response surfaces.* Wiley, New York.

Box, G. E. P. and Wilson, K. B. (1951). On the experimental attainment of optimum conditions. *Journal of the Royal Statistical Society Series* **B 13**, 1–38.

Bradley, R. A. and Yeh, C. M. (1980). Trend-free block designs: Theory. *Annals of Statistics* **8**, 883–893.

Carter, R. L. and Yang, M. C. K. (1986). Large sample inference in random coefficient regression models. *Communications in Statistics – Theory Methods* **15**, 2507–2525.

Chaloner, K. and Verdinelli, I. (1995). Bayesian experimental design : A review. *Statistical Science* **10**, 273-304.

Chernoff, H. (1953). Locally optimal designs for estimating parameters. *Annals of Mathematical Statistics* **24**, 586–602.

de la Garza, A. (1954). Spacing of information in polynomial regression. *Annals of Mathematical Statistics* **25**, 123–130.

Diggle, P. J. (1994). *analysis of longitudinal data.* Oxford: Clarendon Press.

Elfving, G. (1952). Optimum allocation in linear regression theory. *Annals of Mathematical Statistics* **23**, 255–262.

Elfving, G. (1955). Geometric allocation theory. *Skandinavisk Aktuarietidskrift* **37**, 170–190.

Elfving, G. (1956). Selection of nonrepeatable observations for estimation. *Proceedings of the 3rd Berkeley Symposium of Mathematical Statistics and Probability* **1**, 69–75.

Elfving, G. (1959). Design of linear experiments. *Probability and statistics. The Harald Cramér Volume* (ed. by Ulf Grenander), 58–74. Wiley, New York.

Elston, R. C. and Grizzle, J. F. (1962). Estimation of time response curves and their confidence bands. *Biometrics* **18**, 148–159.

Fedorov, V. V. (1972). *Theory of optimal experiments.* Academic Press, New York.

Gaffke, N. and Heiligers, B. (1996). Second order methods for solving extremum problems from optimal linear regression designs. *Optimization* **36**, 41–57.

Ghosh, S. and Rao, C. R. (1996, eds.). *Handbook of Statistics* **13**. Elsevier, Amsterdam.

Grizzle, J. F. and Allen, D. M. (1969). Analysis of growth and dose response curves. *Biometrics* **25**, 357–382.

Guest, P. G. (1958). The spacing of observations in polynomial regression. *Annals of Mathematical Statistics* **29**, 294–299.

Hoel, P. G. (1958). Effiency problems in polynomial estimation. *Annals of Mathematical Statistics* **29**, 1134–1145.

Hoel P. G. (1965). Minimax designs in two dimensional regression. *Annals of Mathematical Statistics.* **36**, 1097–1106.

Jennrich, R. I. and Schluchter, M. D. (1986). Unbalanced repeated measure models with structured covariance matrices. *Biometrics* **42**, 805–820.

Jones, R. H. and Ackerson, C. M. (1990). Serial correlation in equally spaced longitudinal data. *Biometrika* **77**, 721–731.

Jones, R. H. and Boadi-Boateng, F. (1991). Unequally spaced longitudinal data with AR(1) serial correlation. *Biometrics* **47**, 161–175.

Karlin, S. and Studden, W. J. (1966). Optimal experimental designs. *Annals of Mathematical Statistics* **37**, 783–815.

Khatri, C. G. (1966). A note on a MANOVA model applied to problems in growth curves. *Annals of the Institute of Statistical Mathematics* **18**, 75–86.

Kiefer, J. C. (1961). Optimum designs in regression problems, II. *Annals of Mathematical Statistics* **32**, 298–325.

Kiefer, J. C. (1962). An extremum result. *Canadian Journal of Mathematics* **14**, 597–601.

Kiefer, J. C. (1974). General equivalence theory for optimum designs (approximate theory). *Annals of Statistics* **2**, 849–879.

Kiefer, J. C. (1975). Construction and optimality of generalized Youden designs. In *J. N. Srivastava Ed. A Survey of Statistical Design and Linear Models.* North-Holland, Amsterdam, 333–353.

Kiefer, J. C. and Wolfowitz J. (1959). Optimum designs in regression problems. *Annals of Mathematical Statistics* **30**, 271–294.

Kiefer, J. C. and Wolfowitz J. (1960). The eiquivalence of two extremum problems. *Canadian Journal of Mathematics* **12**, 363–366.

Laird, N. M., Lange, N. and Stram, D. (1987). Maximum likelihood computations with repeated measures: Application of the EM algorithm. *Journal of the Amererican Statististical Association* **82**, 97–105.

Laird, N. M. and Ware, J. H. (1982). Random-effects models for longitudinal data. *Biometrics* **38**, 963–974.

Lindley, D. V. (1956). On a measure of the information provided by an experiment. *Annals of Mathematical Statistics* **27**, 986–1005.

Lindsay, J. K. (1993). *Models for repeated measurements.* Oxford: Clarendon Press.

Lindstrom, M. J. and Bates, D. M. (1988). Newton-Rapson and EM algorithms for linear mixed-effects model for repeated-measures data. *Journal of the Amererican Statististical Association* **83**, 1014–1022.

Liski, E. P. and Nummi, T. (1995a). Prediction and inverse estimation in repeated-measures models. *Journal of Statistical Planning and Inference* **47**, 141–151.

Liski, E. P. and Nummi, T. (1995b). Prediction of tree stems to improve efficiency in automatized harvesting of forests. *Scandinavian Journal of Statistics* **22**, 255-259.

Liski, E. P. and Nummi, T. (1996a). The marking for bucking under uncertainty in automatic harvesting of forests. *The International Journal of Production Economics* **46–47**, 373–385.

Liski, E. P. and Nummi, T. (1996b). Prediction in repeated-measures models with engineering applications. *Technometrics* **38**, 25–36.

Liski, E. P., Luoma, A. and Sinha, Bikas K. (1996). Optimal designs in a random coefficient linear growth curve model. *Calcutta Statistical Association Bulletin* **46**, 211–229.

Liski, E. P., Luoma, A., Mandal, N. K. and Sinha, Bikas K. (1998). Optimal designs for prediction in random coefficient linear regression models. *Journal of Combinatorics, Information and System Sciences* (J. N. Srivastava Felicitation Volume), **23**(1–4), 1–16.

Lopes Troya, J. (1982). Optimal designs for covariate models. *Journal of Statistical Planning and Inference* **6**, 373–419.

Minkin, S. (1993). Experimental design for clonogenic assay in chemotherapy. *Journal of the Americal Statistical Association* **88**, 410–420.

Morgan, J. P. (1996). Nested designs. In *S. Ghosh and C. R. Rao (ed.) Handbook of Statistics* **13**, 939–976.

Ozawa, K., Jimbo, M., Kageyama, S. and Mejza, S. (2001). Optimality and construction of incomplete split-block designs. To appear in *Journal of Statistical Planning and Inference.*

Pázman, A. (1986). *Foundations of optimum experimental design.* Reidel, Dordrecht.

Plackett, R. L. and Burman, J. P. (1946). The design of optimum multifactorial experiments. *Biometrika* **33**, 305–325.

Potthoff, R. F. and Roy, S. N. (1964). A generalized multivariate analysis of variance model useful especially for growth curve problems. *Biometrika* **51**, 313–326.

Pukelsheim, F. (1980). On linear regression designs which maxmize information. *Journal of Statistical Planning and Inference* **4**, 339–364.

Pukelsheim, F. (1993). *Optimal design of experiments.* Wiley, New York.

Puntanen, S. and Styan, G. P. H. (1989). The equality of the ordinary least squares estimator and the best linear unbiased estimator. *The American Statistician* **43**(3), 153–163.

Rao, C. R. (1959). Some problems involving linear hypotheses in multivariate analysis. *Biometrika* **46**, 49–58.

Rao, C. R. (1965). The theory of least squares when the parameters are stochastic and its application to the analysis of growth curves. *Biometrika* **52**, 447–458.

Rao, C. R. (1967). Least squares theory using an estimated dispersion matrix and its application to measurement of signals. *Proceedings of the Fifth Berkeley Symposium on Mathematical Statistics and Probability* **1**, 355–372.

Rao, C. R. (1973). Some combinatorial problems of arrays and applications to experimental designs. In *A Survey of Combinatorial Theory* . Eds. J. N. Srivastava, F. Harray, C. R .Rao, G. C. Rota and S. S. Shrikhande. North-Holland, Amsterdam,349–359.

Schwabe, R. (1996). *Optimum designs for multi-factor models.* Lecture Notes in Statistics **113**. Springer, New York.

Searle, S. (1971). *Linear models.* Wiley, New York.

Searle, S. (1987). *Linear models for unbalanced data.* Wiley, New York.

Searle, S. R., Casella, G. and McCulloch, C. E. (1992). *Variance components.* Wiley, New York.

Shah, K. R. and Sinha, Bikas K. (1989). *Theory of optimal designs.* Lecture Notes in Statistics **54**. Springer, New York.

Silvey, S. D. (1978). Optimal design measures with singular information matrices. *Biometrika* **65**, 553–559.

Silvey, S. D. (1980). *Optimum design.* Chapman & Hall, London.

Sinha, Bikas K. (1970). On the Optimality of some designs. *Calcutta Statistical Association Bulletin* **19**, 1–22.

Smith, K. (1918). On the standard deviations of adjusted and interpolated values of an observed polynomial function and its constants and the guidance they give towards a proper choice of the distribution of observations. *Biometrika* **12**, 1–85

Swamy, P. A. V. B. (1971). *Statistical inference in random coefficient regression models.* Lecture Notes in Operational Research and Mathematical Systems **55**. Springer, New York.

Wynn, H. P. (1972). Results in the theory and construction of D-optimum experimental designs. *Journal of the Royal Statistical Society Series* **B 34**, 133–147.

# 2

# Optimal Regression Designs in Symmetric Domains

## Summary


**Features**

**Model(s):** Fixed coefficient regression models
- Single factor polynomial
- Multi-factor linear

**Symmetric experimental domains:** Interval, hypercube and unit ball
**Major tools:** de la Garza (DLG) phenomenon and Loewner order domination of information matrices for search reduction
**Optimality criteria:** Maximization of optimality functionals
**Optimality results:** Specific optimal designs for estimation of regression parameters under continuous design theory
**Thrust:** Symmetry, invariance and concavity of optimality functionals vis - a - vis symmetric experimental domains


In this chapter, some aspects of characterization and construction of optimal regression designs with respect to symmetric experimental domains are discussed. The models assumed are: single factor polynomial regression or multi factor linear regression. The experimental domains are based on the symmetric interval $\mathcal{T} = [-1, 1]$. de la Garza (DLG) phenomenon and Loewner order domination of information matrices are conveniently exploited to obtain a complete class theorem. Further, invariance, concavity and isotonicity of the criterion function are used to further reduce the class of competing designs. Optimal designs for specific optimality criteria are then derived. In the process the roles of Hadamard matrices, complete factorial designs and Caretheodory's Theorem are explained.

## 2.1 Introduction

In this chapter, we propose to undertake a study of Loewner ordering of information matrices in the framework of polynomial regression designs. We start in Section 2.2 with a description of the technique for comparison of two designs by examining if the difference of their information matrices is non-negative definite (nnd). This technique often results in a substantial reduction in the class of competing designs.

In Section 2.3 we deal with polynomial fit models where the space of possible values of the regressor variable is symmetric around the origin. We identify A-, D-, E-, and MV-optimal designs for the model with a quadratic regression function.

Section 2.4 deals with multi-factor first degree polynomial regression models, again requiring that for each factor (i.e., regressor variable) the experimental domain is symmetric about the origin. We discuss optimality results in situations where the space of the regressors is either a Euclidean ball or a unit hypercube.

In both the situations above, symmetry of the space of regressor variables provides a substantial reduction in the class of competing designs. The results of these two sections are not entirely new but are given here to illustrate the technique of Loewner comparison of designs and also for the sake of completeness.

## 2.2 Loewner Comparison of Designs

We say that a design $d_1$ dominates another design $d_2$ in the Loewner sense if $\mathcal{I}_1 - \mathcal{I}_2$ is a non-negative definite (nnd) matrix, where $\mathcal{I}_1$ and $\mathcal{I}_2$ are the information matrices under the designs $d_1$ and $d_2$ respectively. We also denote $\mathcal{I}_1 \geq \mathcal{I}_2$ or $\mathcal{I}_1 - \mathcal{I}_2 \geq 0$ when $\mathcal{I}_1 - \mathcal{I}_2$ is nnd. Thus Loewner partial ordering among information matrices induces a partial ordering among the associated designs. We shall denote $d_1 \succ d_2$, when $d_1$ dominates $d_2$.

It is well-known that within the set of positive definite matrices, matrix inversion is antitonic (or decreasing) with respect to the Loewner ordering i.e., for any two positive definite matrices $\mathbf{A}_1$ and $\mathbf{A}_2$ of the same order, $\mathbf{A}_1 \leq \mathbf{A}_2$ if and only if $\mathbf{A}_1^{-1} \geq \mathbf{A}_2^{-1}$ (vide Kiefer 1959, page 287; Horn and Johnson 1985, page 471). Therefore, Loewner comparison of designs can be based on dispersion matrices of the estimators of the regression parameters or on the corresponding information matrices.

Let $\tilde{\boldsymbol{\beta}}_1$ and $\tilde{\boldsymbol{\beta}}_2$ be the GLS estimators of $\boldsymbol{\beta}$ in (1.3.1) under the competing designs $d_1$ and $d_2$, respectively, and let $\mathbf{X}_1$ and $\mathbf{X}_2$ be the corresponding design matrices in (1.3.1). It is then clear by (1.3.9) and (1.3.10) that

$$\mathbf{V}\left(\tilde{\boldsymbol{\beta}}_1\right) \leq \mathbf{V}\left(\tilde{\boldsymbol{\beta}}_2\right) \Leftrightarrow (\mathbf{X}_1'\mathbf{X}_1)^{-1} \leq (\mathbf{X}_2'\mathbf{X}_2)^{-1} \Leftrightarrow (\mathbf{X}_1'\mathbf{X}_1) \geq (\mathbf{X}_2'\mathbf{X}_2) \tag{2.2.1}$$

where $\left(\mathbf{X}_j'\mathbf{X}_j\right)^{-1}$, $j = 1, 2$ are the dispersion matrices of the OLS estimators of $\boldsymbol{\beta}$ in the ordinary fixed effects regression model (1.3.6). Thus we have proved the following result.

**Theorem 2.2.1** *Let $d_1$ and $d_2$ be designs for the BLU estimator of $\boldsymbol{\beta}$ in the RCR model (1.3.1). A design $d_1$ dominates $d_2$ in the Loewner ordering sense if and only if $d_1$ dominates $d_2$ in the corresponding ordinary regression model (1.3.6).*

In general, there exists no Loewner optimal design $d^*$ such that $d^*$ would dominate any other design $d$ in the Loewner ordering sense (vide Pukelsheim 1993, p. 104).

An *optimality criterion* $\phi$ is a function $\phi$ from the closed cone of nonnegative definite matrices into the real line (vide Shah and Sinha 1989, Chapter 1; Pukelsheim 1993, page 114). An information matrix $\mathcal{I}_1$ is *at least as good as* another information matrix $\mathcal{I}_2$, relative to the criterion $\phi$, when $\phi(\mathcal{I}_1) \geq \phi(\mathcal{I}_2)$. It is essential that a reasonable criterion conforms to the Loewner ordering in the sense that it preserves the matrix ordering (isotonicity),

$$\mathcal{I}_1 \geq \mathcal{I}_2 \geq 0 \Rightarrow \phi(\mathcal{I}_1) \geq \phi(\mathcal{I}_2). \tag{2.2.2}$$

The most prominent optimality criteria are the matrix means $\phi_t$, for $t \in (-\infty; 2]$, which enjoy many desired properties (vide Shah and Sinha 1989, Chapter 1 and Pukelsheim 1993, p. 119 and Chapter 6). The classical A-, D- and E- optimality criteria are special cases of $\phi_t$. Let $\lambda_1, \lambda_2, \ldots, \lambda_p$ denote the eigenvalues of a positive definite information matrix $\mathcal{I}$. Then the matrix mean $\phi_t(\mathcal{I})$ is defined by

$$\phi_t(\mathcal{I}) = \left(\frac{1}{p}\sum_{i=1}^{p} \lambda_i^t\right)^{\frac{1}{t}}. \tag{2.2.3}$$

Because of the formula $\mathbf{V}[\tilde{\boldsymbol{\beta}}(d)] = \mathcal{I}(d)^{-1}$, the matrix mean criteria can be expressed in terms of dispersion matrices as well. Criterion $\phi_{-1}$ is the A-optimality criterion. Maximizing $\phi_{-1}(\mathcal{I})$ is equivalent to minimizing the trace of the corresponding dispersion matrix. The determinant criterion (D-criterion ) $\phi_0(\mathcal{I})$ is equal to $|\mathcal{I}|$, and hence it induces the same preordering among information matrices as $|\mathcal{I}|$. The extreme member of $\phi_t$ for $t \to -\infty$ yields the smallest-eigenvalue criterion (E-criterion) $\phi_{-\infty}(\mathcal{I}) = \lambda_{\min}(\mathcal{I})$.

A criterion function $\phi$ is *superadditive* when

$$\phi(\mathbf{A} + \mathbf{B}) \geq \phi(\mathbf{A}) + \phi(\mathbf{B}) \tag{2.2.4}$$

for all nonnegative definite $\mathbf{A}$ and $\mathbf{B}$. For example the matrix mean $\phi_t$ is superadditive. We may note in passing that $\mathcal{I}_R(d)$, the information matrix for the RCR model (1.3.1) is related to $\mathcal{I}_F(d)$, the information matrix for the fixed regression coefficients model (1.3.6) by the relation

$$[\mathcal{I}_R(d)]^{-1} = \mathbf{D} + [\mathcal{I}_F(d)]^{-1} \tag{2.2.5}$$

The $A$-optimality functional is given by $\phi_{-1}(\mathcal{I})$. It is easy to verify that the $A$-optimality functional for $\mathcal{I}_R$ and $\mathcal{I}_F$ are related by

$$[\phi_{-1}(\mathcal{I}_R)]^{-1} = \frac{1}{p} tr(\mathbf{D}) + [\phi_{-1}(\mathcal{I}_F)]^{-1}, \tag{2.2.6}$$

$p$ being the number of eigenvalues of $\mathcal{I}$. It follows from the above that a design which maximizes $[\phi_{-1}(\mathcal{I}_F)]$ will maximize $[\phi_{-1}(\mathcal{I}_R)]$. The converse is also true. Thus we have established the following result.

**Theorem 2.2.2** *A design $d^*$ is A-optimal for $\boldsymbol{\beta}$ in the RCR model (1.3.1) if and only if it is A-optimal for $\boldsymbol{\beta}$ in the fixed effects regression model (1.3.6).*

**Remark 2.2.1** We are usually referring to maximization of the optimality functional $\phi$. However, there will also be instances where we will be interested in minimization of a suitably defined optimality functional while comparing information matrices.

**Remark 2.2.2** In the rest of this Chapter, we will confine to the regression model (1.3.6) only.

## 2.3 Polynomial Fit Models

We postulate now a polynomial fit model of degree $k \geq 1$,

$$Y_{ij} = \beta_0 + \beta_1 x_i + \cdots + \beta_k x_i^k + e_{ij} \tag{2.3.1}$$

where

$$E(e_{ij}) = 0 \text{ and } V(e_{ij}) = \sigma^2$$

for $i = 1, 2, \ldots, n$ and $j = 1, 2, \ldots, N_i$. The responses $Y_{ij}$ are uncorrelated and the experimental conditions $x_1, x_2, \ldots, x_n$ are assumed to lie in $[-1, 1]$. Note that now the experimental domain $\mathcal{T} = [-1, 1]$ is an interval symmetric with respect to origin. The corresponding regression range $\chi = \{(1, x, \ldots, x^k)' : x \in \mathcal{T}\}$ is a one-dimensional curve embedded in $\mathbf{R}^{k+1}$. We remind ourselves that any collection $d_n = \{x_1, x_2, \ldots, x_n; p_1, p_2, \ldots, p_n\}$ of $n \geq 1$ distinct points $x_i \in \mathcal{T}$ and positive numbers $p_i$, $i = 1, 2, \ldots, n$ such that $\sum_{i=1}^{n} p_i = 1$, induces a continuous design $d$ on the regression range $\chi$ (cf. Pukelsheim 1993, p. 32). In what follows we will denote by $\mathcal{D}$ the set of all such designs. The *exact* design above yields $p_i = N_i/N$ where $N = \sum N_i$ is the total number of observations.

In subsection 1.2.1, we stated the well-known de la Garza (1954) phenomenon. Let $d_n = \{x_1, x_2, \ldots, x_n;\ p_1, p_2, \ldots, p_n\}$ with $n > k + 1$ be an $n$-point design for the LSE of $\boldsymbol{\beta} = (\beta_0, \beta_1, \ldots, \beta_k)'$ in the polynomial fit model (2.3.1) of degree $k$. Then there exists a $(k + 1)$-point design $d_{k+1}^* = \{x_1^*, x_2^*, \ldots, x_{k+1}^*;\ p_1^*, p_2^*, \ldots, p_{k+1}^*\}$ for the LSE of $\boldsymbol{\beta}$ in (2.3.1) such that $\mathcal{I}(d_{k+1}^*) = \mathcal{I}(d_n)$, where $\mathcal{I}(d_n)$ denotes the information matrix of the design $d_n$.

### 2.3.1 Symmetric Polynomial Designs

First we consider the *reflection* operation. Let $d \in \mathcal{D}$ be a design for the LSE of $\boldsymbol{\beta} = (\beta_0, \beta_1, \ldots, \beta_k)'$ on $\mathcal{T} = [-1, 1]$ in the polynomial fit model (2.3.1). The reflected design $d^R$ is given by $d^R = \{-x_1, -x_2, \ldots, -x_n; p_1, p_2, \ldots, p_n\}$. The designs $d$ and $d^R$ have the same even moments, while the odd moments of $d^R$ have a reversed sign.

If $\mathcal{I}(d^R)$ denotes the $(k+1) \times (k+1)$ information matrix of $d^R$, then

$$\mathcal{I}(d^R) = \mathbf{Q}\mathcal{I}(d)\mathbf{Q}, \tag{2.3.2}$$

where $\mathbf{Q} = \text{Diag}(1, -1, 1, -1, \ldots, \pm 1)$ is a diagonal matrix with diagonal elements 1, $-1$, 1, $-1$, ..., $\pm 1$.

The symmetrized design

$$\bar{d} = \frac{1}{2}(d + d^R) \equiv \{\pm x_i;\ \frac{p_i}{2}, \frac{p_i}{2} | 1 \le i \le n\}$$

assigns the weights $\frac{p_i}{2}$ to $x_i$ and $-x_i$ for each $i$. The information matrix of $\bar{d}$ is

$$\mathcal{I}(\bar{d}) = \frac{1}{2}[\mathcal{I}(d) + \mathbf{Q}\mathcal{I}(d)\mathbf{Q}],$$

where all odd moments are zero and the even moments are equal to the corresponding moments of the original design $d$. Hence the averaging operation simplifies information matrices by letting all odd moments vanish. Since $\mathcal{I}(d^R)$ is obtained from $\mathcal{I}(d)$ by the similarity transformation (2.3.2), $\mathcal{I}(d^R)$ and $\mathcal{I}(d)$ have the same eigenvalues.

From the above, it follows that any optimality criterion which is a function of the eigenvalues of the information matrices will be invariant with respect to the reflection operation. We refer to Section 2.2 for a discussion on Loewener comparison of designs. It follows that superadditivity and invariance of $\phi$ (with respect to the reflection) imply

$$\begin{aligned}
\phi[\mathcal{I}(\bar{d})] &= \phi\left\{\frac{1}{2}[\mathcal{I}(d) + \mathcal{I}(d^R]\right\} \\
&\ge \frac{1}{2}\{\phi[\mathcal{I}(d)] + \phi[\mathcal{I}(d^R)]\} \\
&= \phi[\mathcal{I}(d)].
\end{aligned}$$

Thus symmetrization improves the value of the criterion $\phi$, or at least maintains the same value, provided that $\phi$ is superadditive and invariant with respect to the reflection. Therefore, for such criteria, we may confine ourselves to the class of symmetric designs.

### 2.3.2 Symmetric Designs for Quadratic Regression

Let $d_n$ denote a symmetric n-point design on $\mathcal{T} = [-1, 1]$ for the LSE of $\boldsymbol{\beta} = (\beta_0, \beta_1, \beta_2\}'$ in the quadratic regression model

$$Y_{ij} = \beta_0 + \beta_1 x_i + \beta_2 x_i^2 + e_{ij} \tag{2.3.3}$$

with the assumptions similar to the general polynomial fit model (1.2.1). It follows that

$$\mathcal{I}(d_n) = \begin{pmatrix} 1 & 0 & \mu_2 \\ 0 & \mu_2 & 0 \\ \mu_2 & 0 & \mu_4 \end{pmatrix}.$$

Whenever $n > 3$, we may obtain the *same* information matrix by using $d_3 = \{-a, 0, a; w/2, 1-w, w/2\}$ by choosing $a = \sqrt{\mu_4/\mu_2}$ and $w = \mu_2^2/\mu_4$. This is the spirit of de la Garza (DLG) Phenomenon. Henceforth, we will confine only to 3-point symmetric designs.

Now we prove for the model (2.3.3) the following result.

**Theorem 2.3.1** *Let $d_3$ be any 3-point symmetric design for the LSE of $\beta = (\beta_0, \beta_1, \beta_2)'$ in (2.3.3) with its support points in the interior of $\mathcal{T}$. Then there exists a symmetric 3-point design $d_3^* = \{-1, 0, 1;\ \frac{p}{2}, 1-p, \frac{p}{2}\}$ with $p < 1$ such that $d_3^* \succ d_3$ in the sense that $\mathcal{I}(d_3^*) \geq \mathcal{I}(d_3)$.*

**Proof.** We start with $d_3$ as defined before and examine the nature of $d_3^*$ as against $d_3$. By definition

$$\mathcal{I}(d_3^*) - \mathcal{I}(d_3) = \begin{pmatrix} 1 & 0 & p \\ 0 & p & 0 \\ p & 0 & p \end{pmatrix} - \begin{pmatrix} 1 & 0 & a^2w \\ 0 & a^2w & 0 \\ a^2w & 0 & a^4w \end{pmatrix}. \qquad (2.3.4)$$

For the choice $p = a^2w$ all elements of the difference matrix (2.3.4) are zero, except the last diagonal element $p - a^4w = a^2(1-a^2)w$, which is positive since $0 < a < 1$ and $0 < w < 1$. Thus $\mathcal{I}(d_3^*) - \mathcal{I}(d_3)$ is nonnegative definite for $p = a^2w$. Thus our proof is complete. □

**Remark 2.3.1** As a matter of fact, Kiefer (1959) established a more general result which can be stated as follows in the present set-up of quadratic regression. Let $d_3$ be any arbitrary (not necessarily symmetric) 3-point design whose support points do not include both the extremes $\pm 1$. Then there exists a 3-point design say, $d_3^*$ which includes both the extreme points and a suitable interior point with appropriate weights for which $\mathcal{I}(d_3^*) \geq \mathcal{I}(d_3)$ holds in the LOD sense. For the sake of completeness, we will provide an explicit proof of a certain version of this result. This will be given in Section 3.2. This shows that the class of 3-point designs of the type $d_3^*$ constitute the class of admissible designs irrespective of the nature of the optimality functional $\phi$.

Now the A-,D-, E- and MV-optimal designs can be determined straightforwardly. Note that in terms of matrix means $\phi_t$, $-\infty \leq t \leq 1$, $\phi_{-1}$, $\phi_0$ and $\phi_{-\infty}$ are the A-, D- and E-optimality criteria, respectively. The characteristic function of $\mathcal{I}(d_p)$ corresponding to a 3-point design $d_p$ is

$$f(\lambda) = (p - \lambda)[(1-\lambda)(p-\lambda) - p],$$

which yields the eigenvalues of $\mathcal{I}(d_p)$:

$$\begin{aligned}\lambda_1 &= \frac{1}{2}(p+1)+\frac{1}{2}\sqrt{5p^2-2p+1},\\ \lambda_2 &= p,\\ \lambda_3 &= \frac{1}{2}(p+1)-\frac{1}{2}\sqrt{5p^2-2p+1}.\end{aligned}$$

The optimal designs are as follows:

$$\text{A-optimal design: } \left\{-1,0,1;\ \frac{1}{4},\frac{1}{2},\frac{1}{4}\right\},$$

$$\text{D-optimal design: } \left\{-1,0,1;\ \frac{1}{3},\frac{1}{3},\frac{1}{3}\right\},$$

$$\text{E-optimal design: } \left\{-1,0,1;\ \frac{1}{5},\frac{3}{5},\frac{1}{5}\right\}.$$

$$\text{MV-optimal design: } \left\{-1,0,1;\ \frac{1}{4},\frac{1}{2},\frac{1}{4}\right\}.$$

In the above, MV-optimal design is based on the criterion of minimization of the larger of the two variances of the estimators involving the linear and the quadratic terms.

It is clear from the nature of the above designs that, within the class of 3-point designs, further Loewner order domination is *not* possible. So there exists no Loewner optimal design. In spite of this limitation, it would be natural to examine the nature of a complete class of designs for a general polynomial regression model. Pukelsheim's (1993) Claim 10.7 states that $d$ is admissible in (2.3.1) if and only if $d$ has at most $k-1$ support points in the open interval $(-1,1)$. Thus the $(k+1)$-point designs $d_{k+1} = \{-1, t_2, \ldots, t_s, 1; p_1, p_2, \ldots, p_s, p_{k+1}\}$ with $t_2$, $t_3$, ..., $t_k \in (-1,1)$ and $\sum_{i=1}^{s+1} p_i = 1$ are admissible. This gives a starting point to look for spesific optimal designs. Pukelsheim (1993) has listed A-, E- and D-optimal designs for polynomial regression from the 1st to the 10th degree over $[-1,1]$. The theoretical developments leading to such computations are all explained in Pukelsheim (1993).

## 2.4 Multi-Factor First-Degree Polynomial Fit Models

Let us first look at an $m$-factor first-degree polynomial fit model

$$Y_{ij} = \beta_1 x_{i1} + \beta_2 x_{i2} + \cdots + \beta_m x_{im} + e_{ij} \tag{2.4.1}$$

with $m$ regressor variables, $n$ experimental conditions $\mathbf{x}_i = (x_{i1}, x_{i2}, \ldots, x_{im})'$, $i = 1, 2, \ldots, n$ ; $j = 1, 2, \ldots, N_i$ and the model has no constant term.

In polynomial fit models of the previous section the experimental domain is $\mathcal{T} = [-1, 1]$.

For an $m$-way polynomial fit model (2.4.1) the experimental domain $\mathcal{T}$ is a subset of the $m$-dimensional Euclidean space $\mathbf{R}^m$. In this section we consider two extensions of the one-dimensional domain $\mathcal{T} = [-1, 1]$: A Euclidean ball of radius $\sqrt{m}$ and a symmetric $m$-dimensional hypercube $\mathcal{T} = [-1, 1]^m$ with half of sidelength 1. Later in this section we also study an $m$-way first-degree polynomial fit model with a constant term.

### 2.4.1 Designs in a Euclidean Ball

We assume now that the experimental domain for the model (2.4.1) is an $m$-dimensional Euclidean ball of radius $\sqrt{m}$, that is $\mathcal{T}_{\sqrt{m}} = \{\mathbf{x} \in \mathbf{R}^m : ||\mathbf{x}|| \leq \sqrt{m}\}$, where $||\cdot||$ denotes the Euclidean norm.

Denote $\mu_{jk} = \sum_{i=1}^{n} p_i x_{ij} x_{ik}$ for $j$, $k = 1, 2, \ldots, m$. Then the information matrix of an $n$-point design

$$d = \{\mathbf{x}_1, \mathbf{x}_2, \ldots, \mathbf{x}_n;\ p_1, p_2, \ldots, p_n\} \tag{2.4.2}$$

is of the form

$$\mathcal{I}(d) = \sum_{i=1}^{n} p_i \mathbf{x}_i \mathbf{x}_i' = \begin{pmatrix} \mu_{11} & \mu_{12} & \cdots & \mu_{1m} \\ \mu_{21} & \mu_{22} & \cdots & \mu_{2m} \\ \vdots & \vdots & \ddots & \vdots \\ \mu_{m1} & \mu_{m2} & \cdots & \mu_{mm} \end{pmatrix}. \tag{2.4.3}$$

Consider an $n$-point design (2.4.2), $n \geq m$. Let

$$\lambda_1 \mathbf{w}_1 \mathbf{w}_1' + \lambda_2 \mathbf{w}_2 \mathbf{w}_2' + \cdots + \lambda_m \mathbf{w}_m \mathbf{w}_m' = \mathcal{I}(d)$$

be the spectral decomposition of $\mathcal{I}(d)$, where $\mathbf{w}_i$ and $\lambda_i$ ($> 0$) are orthonormal eigenvectors and the eigenvalues of $\mathcal{I}(d)$, respectively. Note that

$$\begin{aligned} tr[\mathcal{I}(d)] &= \sum_{i=1}^{m+1} \lambda_i \\ &= \sum_{i=1}^{n} p_i \mathbf{x}_i' \mathbf{x}_i \leq m, \end{aligned}$$

since by assumption $\mathbf{x}_i \in \mathcal{T}_{\sqrt{m}}$ for all $i = 1, 2, \ldots, n$.

Denote $\widetilde{\mathbf{w}}_i = \sqrt{\sum_{i=1}^{n} \lambda_i} \mathbf{w}_i$ and $r_i = \frac{\lambda_i}{\sum_{i=1}^{n} \lambda_i}$, $i = 1, 2, \ldots, m$ and consider the $m$-point design

$$\widetilde{d} = \{\widetilde{\mathbf{w}}_1, \widetilde{\mathbf{w}}_2, \ldots, \widetilde{\mathbf{w}}_m;\ r_1, r_2, \ldots, r_m\}.$$

Clearly, $\widetilde{\mathbf{w}}_i \in \mathcal{T}_{\sqrt{m}}$, $i = 1, 2, \ldots, m$, and the designs $\widetilde{d}$ and $d$ have the same information matrix, i.e. $\widetilde{d}$ and $d$ are information equivalent designs. Thus for any $n$-point design $d$ for the LSE of $\boldsymbol{\beta}$ in (2.4.1) there exists an information equivalent $m$-point design $\tilde{d}$ from the regression range $\mathcal{T}_{\sqrt{m}}$ such that the support vectors are orthogonal. We say that $\tilde{d}$ is an *orthogonal design.* This incidentally demonstrates validity of the DLG phenomenon in the present set-up as well.

Now we will prove that any design $d$ for the LSE of $\boldsymbol{\beta} = (\beta_1, \beta_2, \ldots, \beta_m)'$ in (2.4.1) can be *dominated* by a suitably defined orthogonal design as well. Hence an optimal design, if it exists, is necessarily an orthogonal design.

**Theorem 2.4.1** *Let $d$ be an $n$-point design for the LSE of $\boldsymbol{\beta} = (\beta_1, \beta_2, \ldots, \beta_m)'$ in (2.4.1), $n \geq m$. Then there exists an $m$-point orthogonal design $\widehat{d}$ that dominates $d$ in the Loewner sense, i.e. $\widehat{d} \succ d$.*

**Proof.** Let any $n$-point design $d$ with $n \geq m$ be given. Then there exists, as shown above, an information equivalent $m$-point design $\tilde{d} = \{\mathbf{v}_1, \mathbf{v}_2, \ldots, \mathbf{v}_m; r_1, r_2, \ldots, r_m\}$ with orthogonal support vectors $\mathbf{v}_i \in \mathcal{T}_{\sqrt{m}}$, $i = 1, 2, \ldots, m$. Then we can always define such an $m$-point design $\widehat{d} = \{\widehat{\mathbf{v}}_1, \widehat{\mathbf{v}}_2, \ldots \widehat{\mathbf{v}}_m; r_1, r_2, \ldots r_m\}$ that $\widehat{\mathbf{v}}_i = \sqrt{m}\mathbf{v}_i/||\mathbf{v}_i||$. Note that $\widehat{d}$ is an orthogonal design *on the ball* $\mathcal{T}_{\sqrt{m}}$.

Next, we prove that $\widehat{d}$ dominates $d$ in the Loewner sense. Since $||\mathbf{v}_i||/\sqrt{m} \leq 1$ for $i = 1, 2, \ldots, m$, we have

$$\mathcal{I}(\widehat{d}) - \mathcal{I}(\tilde{d}) = \sum_{i=1}^{m} r_i \left(1 - \frac{||\mathbf{v}_i||^2}{m}\right) \widehat{\mathbf{v}}_i \widehat{\mathbf{v}}_i' \geq 0, \tag{2.4.4}$$

and consequently $\widehat{d} \succ \tilde{d}$. Since $\mathcal{I}(d) = \mathcal{I}(\tilde{d})$, we have $\mathcal{I}(\widehat{d}) \geq \mathcal{I}(d)$. This concludes our proof. □

It is again clear that there is no Loewner optimal $m$-point orthogonal design for the LSE of $\boldsymbol{\beta}$ in (2.4.1) (This was mentioned earlier immediately after Theorem 2.2.1).

We consider now the $m$-factor first-degree polynomial fit model

$$Y_{ij} = \beta_0 + \beta_1 x_{i1} + \cdots + \beta_m x_{im} + e_{ij}, \quad i = 1, 2, \ldots, n; \ j = 1, 2, \ldots, N_i \tag{2.4.5}$$

with a constant term $\beta_0$. The regression range of an $n$-point design $d$ for LSE of $\boldsymbol{\beta}$ in (2.4.5) is of the form

$$\mathcal{X} = \left\{ \begin{pmatrix} 1 \\ \mathbf{x} \end{pmatrix} \mid \mathbf{x} \in \mathcal{T}_{\sqrt{m}} \right\} \subset \mathcal{T}_{\sqrt{m+1}}. \tag{2.4.6}$$

The information matrix of an $n$-point design $d$ in (2.4.2) is of the form

$$\mathcal{I}(d) = \sum_{i=1}^{n} p_i \begin{pmatrix} 1 \\ \mathbf{x}_i \end{pmatrix} (1, \mathbf{x}_i') = \begin{pmatrix} 1 & \mu_{01} & \cdots & \mu_{0m} \\ \mu_{10} & \mu_{11} & \cdots & \mu_{1m} \\ \vdots & \vdots & \ddots & \vdots \\ \mu_{m0} & \mu_{m1} & \cdots & \mu_{mm} \end{pmatrix}, \tag{2.4.7}$$

where, additionally, $\mu_{0k} = \sum_{i=1}^{n} p_i x_{ik}$, $k = 1, 2, \ldots, m$.

Consider an $n$-point design $d$, $n \geq m+1$. Let

$$\lambda_1 \mathbf{w}_1 \mathbf{w}_1' + \lambda_2 \mathbf{w}_2 \mathbf{w}_2' + \cdots + \lambda_{m+1} \mathbf{w}_{m+1} \mathbf{w}_{m+1}' = \mathcal{I}(d)$$

be the spectral decomposition of $\mathcal{I}(d)$, where $\mathbf{w}_i$ and $\lambda_i > 0$ are orthonormal eigenvectors and the eigenvalues of $\mathcal{I}(d)$ respectively. Note that

$$\begin{aligned} tr[\mathcal{I}(d)] &= \sum_{i=1}^{m+1} \lambda_i \\ &= \sum_{i=1}^{n} p_i(1 + \mathbf{x}_i'\mathbf{x}_i) \leq m+1, \end{aligned}$$

since by assumption $\mathbf{x}_i \in \mathcal{T}_{\sqrt{m}}$ for all $i = 1, 2, \ldots, n$.

Denote $\tilde{\mathbf{w}}_i = \sqrt{\sum_{i=1}^{m+1} \lambda_i \mathbf{w}_i}$ and $r_i = \frac{\lambda_i}{\sum_{i=1}^{m+1} \lambda_i}$, $i = 1, 2, \ldots, m+1$, and consider the $(m+1)$-point design

$$\tilde{d} = \{\tilde{\mathbf{w}}_1, \tilde{\mathbf{w}}_2, \ldots, \tilde{\mathbf{w}}_{m+1};\ r_1, r_2, \ldots, r_{m+1}\}.$$

Clearly, the support vectors $\tilde{\mathbf{w}} \in \mathcal{T}_{\sqrt{m+1}}$, $i = 1, 2, \ldots, m+1$ are orthognal, and the designs $\tilde{d}$ and $d$ have the same information matrix, i.e. $\tilde{d}$ and $d$ are information equivalent designs. Thus, for any $n$-point design $d$ from the regression range (2.4.6) for the LSE of $\boldsymbol{\beta}$ in (2.4.5) there exists an information equivalent $(m+1)$-point design $\tilde{d}$ from the regression range $\mathcal{T}_{\sqrt{m+1}}$ such that the support vectors are orthogonal, i.e. $\tilde{d}$ is an *orthogonal design.*

If the vectors $\mathbf{x}_i \in \mathcal{T}_{\sqrt{m}}$, $i = 1, 2, \ldots, m+1$ fulfill the conditions

$$1 + \mathbf{x}_i'\mathbf{x}_i = 1 + m, \quad 1 + \mathbf{x}_i'\mathbf{x}_j = 0 \qquad (2.4.8)$$

for all $1 \leq i \neq j \leq m+1$, then the vectors span a convex body in $\mathbf{R}^m$ called a *regular simplex* (cf. Pukelsheim 1993, p. 391). A design $d = \{\mathbf{x}_1, \mathbf{x}_2, \ldots, \mathbf{x}_{m+1};\ p_1, p_2, \ldots, p_{m+1}\}$ which places weights $p_i$, $i = 1, 2, \ldots, m+1$, on the vertices of a regular simplex in $\mathbf{R}^m$ is called a *simplex design.* A design with equal weights $p_1 = p_2 = \ldots = p_{m+1} = \frac{1}{m+1}$ is called a *uniform simplex design.* If the vectors $\mathbf{x}_1$, $\mathbf{x}_2$, ..., $\mathbf{x}_{m+1}$ satisfy the conditions (2.4.8), then the vectors $(1, \mathbf{x}_1')$, $(1, \mathbf{x}_2')$, ..., $(1, \mathbf{x}_{m+1}')$ are orthogonal, they belong to the boundary of $\mathcal{T}_{m+1}$ and are of the form (2.4.6). Given an orthogonal design on a Euclidean ball (the support vectors belong to the boundary), then any other orthogonal design can be obtained from it by orthogonal rotation of support vectors.

It is obvious that we can always find $m+1$ orthogonal vectors $(1, \mathbf{x}_i')'$, $i = 1, 2, \ldots, m+1$ such that every $\mathbf{x}_i$ belongs to the boundary of $\mathcal{T}_{\sqrt{m}}$. For $m = 1$ the support points are $x_1 = 1$ and $x_2 = -1$ so that $(1, 1)'$

and $(1,-1)'$ satisfy the conditions (2.4.8). The support points $\mathbf{x}_1$, $\mathbf{x}_2$, $\mathbf{x}_3$ satisfying the conditions (2.4.8) belong to the boundary of $\mathcal{T}_{\sqrt{2}}$ and they span an *equilateral triangle* on the sphere $\mathcal{T}_{\sqrt{2}}$. For example the support points $(1,1)'$, $-\frac{1}{2}(1+\sqrt{3},1-\sqrt{3})'$, $-\frac{1}{2}(1-\sqrt{3},1+\sqrt{3})'$, and every rotation of them span an equilateral triangle.

We consider now design optimality criteria $\phi$ which are isotonic with respect to the Loewner ordering. We prove that any design $d$ for the LSE of $\boldsymbol{\beta}$ in (2.4.5) can be dominated by an $(m+1)$-point simplex design. Hence an optimal design with respect to $\phi$, if it exists, can be found among the $(m+1)$-point simplex designs.

**Theorem 2.4.2** *Let $d$ be an $n$-point design for the LSE of $\boldsymbol{\beta}$ in (2.4.5) over the ball $\mathcal{T}_{\sqrt{m+1}}$, $n \geq m+1$ and let $\phi$ be any optimality criterion that is (1) isotonic with respect to the Loewner ordering and (2) depends on the information matrix only through its eigenvalues. Then there exists an $(m+1)$-point simplex design $d^*$ that dominates $d$ with respect to $\phi$, i.e. $\phi(\mathcal{I}(d^*)) \geq \phi(\mathcal{I}(d))$.*

**Proof.** Let any $n$-point design $d$ with $n \geq m+1$ be given. Then there exists as shown above, an information equivalent $(m+1)$-point design $\tilde{d} = \{\mathbf{w}_1, \mathbf{w}_2, \ldots, \mathbf{w}_{m+1}; r_1, r_2, \ldots, r_{m+1}\}$ with orthogonal support vectors $\mathbf{w}_i \in \mathcal{T}_{\sqrt{m+1}}$, $i = 1, 2, \ldots, m+1$. Then we can always define such an $(m+1)$-point design $d_u = \{\mathbf{u}_1, \mathbf{u}_2, \ldots, \mathbf{u}_{m+1};\ r_1, r_2, \ldots, r_{m+1}\}$ that $\mathbf{u}_i = \sqrt{m+1}\mathbf{w}_i/\|\mathbf{w}_i\|$. Now $d_u$ is an orthogonal design on the ball $\mathcal{T}_{\sqrt{m+1}}$. We note that the range of $d_u$ is *not* of the form (2.4.6).

We have noted earlier that we can find $m+1$ orthogonal vectors $(1,\ \mathbf{x}_i^{'})'$, $i = 1, 2, \ldots, m+1$ such that every $\mathbf{x}_i$ belongs to the boundary of $\mathcal{T}_{\sqrt{m}}$. Let $\widehat{\mathbf{u}}_i = (1,\ \mathbf{x}_i^{'})'$ for every $i$. It then follows that

$$\widehat{\mathbf{u}}_i'\widehat{\mathbf{u}}_i = 1 + \widehat{\mathbf{x}}_i'\widehat{\mathbf{x}}_i = 1 + m, \quad \widehat{\mathbf{u}}_i'\widehat{\mathbf{u}}_j = 1 + \widehat{\mathbf{x}}_i'\widehat{\mathbf{x}}_j = 0$$

Thus $d^* = \{\widehat{\mathbf{u}}_1,\ \widehat{\mathbf{u}}_2,\ \ldots, \widehat{\mathbf{u}}_{m+1};\ r_1,\ r_2,\ \ldots, r_{m+1}\}$ is an orthogonal design on the ball $\mathcal{T}_{\sqrt{m+1}}$. Since any other orthogonal design on the ball can be obtained by an orthogonal rotation of $d^*$, we can find an $(m+1)\times(m+1)$ orthogonal matrix $\mathbf{P}$ such that $\widehat{\mathbf{u}}_i = \mathbf{P}\mathbf{u}_i$ for each $i$. Clearly, the range of $d^*$ is of the form given by (2.4.6). We note that $d^*$ is a regular simplex design.

Next we prove that $d^*$ dominates $d$ with respect to criterion $\phi$. The argument used to prove (2.4.4) shows that $\mathcal{I}(d_u) \geq \mathcal{I}(\tilde{d}) = \mathcal{I}(d)$. Further, $\mathcal{I}(d_u) = \mathbf{P}'\mathcal{I}(d^*)\mathbf{P}$ and hence, $\mathcal{I}(d_u)$ and $\mathcal{I}(d^*)$ have the same eigenvalues. Thus by the assumption (2), $\phi(\mathcal{I}(d^*)) = \phi(\mathcal{I}(d_u))$. It is now clear that $d^*$ dominates $d$ with respect to $\phi$. We note that the range of $d^*$ is of the form (2.4.6) and hence $d^*$ is a valid design for the model given by (2.4.5). Hence the proof is complete. □

Note that the most prominent optimality criteria, the matrix means $\phi_t$, fulfil the conditions of Theorem 2.4.2. Also the DS-optimality criterion, which will be discussed in Chapter 5, is a special case of the above theorem. A slightly more general formulation of Theorem 2.4.2 was proved by Liski and Zaigraev (2001, Theorem 3).

Let $\mathbf{X}$ denote a model matrix whose rows are the support vectors $(1\ \mathbf{x}_1')'$, $(1\ \mathbf{x}_2')', \ldots, (1\ \mathbf{x}_{m+1}')'$ of a simplex design $d^{(m+1)}$. For such a design

$$\mathcal{I}(d^{(m+1)}) = \mathbf{X}'\mathbf{D}\mathbf{X}, \quad \mathbf{D} = \mathrm{Diag}(p_1, p_2, \cdots, p_{m+1})$$

and the model matrix $\mathbf{X}$ is square. The non-zero eigenvalues of $\mathbf{X}'\mathbf{D}\mathbf{X}$ and $\mathbf{D}\mathbf{X}\mathbf{X}'$ are the same and

$$\mathbf{D}\mathbf{X}\mathbf{X}' = (m+1)\mathbf{D} = (m+1)\,\mathrm{Diag}(p_1, p_2, \cdots, p_{m+1}).$$

Thus optimum designs with respect to matrix mean criteria are easy to determine. A design with equal weights $p_1 = p_2 = \cdots = p_{m+1} = 1/(m+1)$, called a *uniform simplex design*, is A- , D- and E-optimal.

### 2.4.2 Designs in a Unit Hypercube

A symmetric $m$-dimensional unit-cube $[-1,1]^m$ is a natural extension of $[-1,1]$. Note that $[-1,1]^m$ is the convex hull of its extreme points, the $2^m$ vertices of $[-1,1]^m$. It is known that in order to find optimal support points, we need to search the extreme points of the regression range $\chi$ only. If the support of a design contains other than extreme points, then it can be Loewner dominated by a design with extreme support points only. This result was basically presented by Elfving (1952, 1959). A unified general theory is given by Pukelsheim (1993, Chapter 8).

As an example consider a 2-factor first degree model (2.4.1) that has *no* constant term. The experimental domain $\mathcal{T}$ is the square $[-1,1]^2$. The extreme points (vertices) of $[-1,1]^2$ are

$$\begin{pmatrix}1\\1\end{pmatrix}, \begin{pmatrix}1\\-1\end{pmatrix}, \begin{pmatrix}-1\\1\end{pmatrix} \text{ and } \begin{pmatrix}-1\\-1\end{pmatrix}.$$

Suppose now that the support of a design $d$ consists of the extreme points only. Then the information matrix of $d$ takes finally the form

$$\mathcal{I}(d) = p\begin{pmatrix}1&1\\1&1\end{pmatrix} + (1-p)\begin{pmatrix}1&-1\\-1&1\end{pmatrix} = \begin{pmatrix}1&2p-1\\2p-1&1\end{pmatrix} \qquad (2.4.9)$$

with $0 < p < 1$, since $\begin{pmatrix}1\\1\end{pmatrix}(1\ \ 1) = \begin{pmatrix}-1\\-1\end{pmatrix}(-1\ \ -1)$, $\begin{pmatrix}1\\-1\end{pmatrix}(1\ \ -1) = \begin{pmatrix}-1\\1\end{pmatrix}(-1\ \ 1)$. The eigenvalues of $\mathcal{I}(d)$ are $\lambda_{1,2} = 1 \pm (2p-1)$. Thus the optimum values of various matrix means criteria $\phi_t$ are easy to determine. The 2-point design

$$d_{\frac{1}{2}}^{(2)} = \left\{\begin{pmatrix}1\\1\end{pmatrix}, \begin{pmatrix}-1\\1\end{pmatrix};\ \frac{1}{2}\right\} \qquad (2.4.10)$$

is A-, D- and E-optimal.

Note that any extreme point design with 2, 3 or 4 support points such that the total weight at the points $\begin{pmatrix}1\\1\end{pmatrix}$ and $\begin{pmatrix}-1\\-1\end{pmatrix}$ is equal to $p$ yields the same information matrix (2.4.9), i.e. they constitute a class of information equivalent designs. The information equivalent 2-point designs with the support $\left\{\begin{pmatrix}1\\1\end{pmatrix},\begin{pmatrix}1\\-1\end{pmatrix}\right\}$, $\left\{\begin{pmatrix}1\\1\end{pmatrix},\begin{pmatrix}-1\\1\end{pmatrix}\right\}$, $\left\{\begin{pmatrix}-1\\-1\end{pmatrix},\begin{pmatrix}1\\-1\end{pmatrix}\right\}$ and $\left\{\begin{pmatrix}-1\\-1\end{pmatrix},\begin{pmatrix}-1\\1\end{pmatrix}\right\}$ have the minimal support size. For example, a 3-point design $\left\{\begin{pmatrix}1\\1\end{pmatrix},\begin{pmatrix}1\\-1\end{pmatrix},\begin{pmatrix}-1\\1\end{pmatrix};\ \frac{1}{2},p_2,p_3\right\}$ with $p_2+p_3=\frac{1}{2}$ is information equivalent to (2.4.10), and hence also it is A-, D- and E-optimal.

Now consider the model (2.4.5) with $m=2$. The information matrix of $d$ defined below for the LSE of $\boldsymbol{\beta}$ in (2.4.5) is

$$\mathcal{I}(d)=\begin{pmatrix}1 & p_1+p_2-p_3-p_4 & p_1-p_2-p_3+p_4\\ p_1+p_2-p_3-p_4 & 1 & p_1+p_2-p_3-p_4\\ p_1-p_2-p_3+p_4 & p_1+p_2-p_3-p_4 & 1\end{pmatrix} \tag{2.4.11}$$

when the support of $d$ is $\left\{\begin{pmatrix}1\\1\\1\end{pmatrix},\begin{pmatrix}1\\1\\-1\end{pmatrix},\begin{pmatrix}1\\-1\\-1\end{pmatrix},\begin{pmatrix}1\\-1\\1\end{pmatrix}\right\}$ and $p_i>0$, $i=1,\ 2,\ 3,\ 4$ with $p_1+p_2+p_3+p_4=1$ are the corresponding weights. It may be noted that it is enough to concentrate on the above extreme points (in the suport of $d$) even for a model *with* the intercept term.

Let $\lambda_1$, $\lambda_2$, $\lambda_3$ denote the eigenvalues of the information matrix (2.4.11). Then by Hadamard's inequality (Horn and Johnson 1985, p. 477)

$$|\mathcal{I}(d)|=\lambda_1\lambda_1\lambda_2\leq 1. \tag{2.4.12}$$

Equality holds in (2.4.12) if and only if $\mathcal{I}(d)$ is equal to $\mathbf{I}_3$, and $\mathcal{I}(d)=\mathbf{I}_3$ exactly when $p_1=p_2=p_3=p_4=\frac{1}{4}$. Consequently, the design that assigns uniform weights $\frac{1}{4}$ to each of these four extreme points of the regression range is the D-optimal design. Since $\lambda_1+\lambda_2+\lambda_3=1$, the design is also A- and E-optimal design. There is no 3-point design $d^{(3)}=\{\mathbf{x}_1,\mathbf{x}_2,\mathbf{x}_3;\ p_1,p_2,p_3\}$ such that $\mathcal{I}(d^{(3)})=\mathbf{I}_3$. This is easy to show if the support is chosen to be any 3-point subset of the vertices of the regression range $\chi$. Therefore a 3-point design cannot be D-, A-or E-optimal.

We now turn to the general case for arbitrary but fixed $m(>2)$. Note that the *complete factorial design* which assigns equal weight $2^{-m}$ to each of the $2^m$ extreme points of the form $\{\pm1,\ \pm1,\ldots,\pm1\}$ provides an information matrix exactly equal to $\mathcal{I}_{m+1}$. Below we establish optimality of such designs.

**Theorem 2.4.3** *Let $\mathcal{D}_n$ be the set of designs $d$ with support size $n \geq m+1$ in the $m$-way first-degree model (2.4.5) on the symmetric unit hypercube $[-1,1]^m$. Then a design $d \in \mathcal{D}_n$ is D-, A- and E-optimal if and only if $\mathcal{I}(d) = \mathbf{I}_{m+1}$.*

**Proof.** The $n \times (m+1)$ model matrix $\mathbf{X}$ has entries 1 and $-1$ only, since we need to search solely the designs whose support consists of the extreme points of the regression range $\chi$. For such a design

$$\mathcal{I}(d) = \mathbf{X}'\mathbf{D}\mathbf{X}, \quad \mathbf{D} = \mathrm{Diag}(p_1, p_2, \dots, p_n)$$

and

$$|\mathcal{I}(d)| = \lambda_1 \lambda_2 \cdots \lambda_{m+1} \leq 1. \tag{2.4.13}$$

The inequality (2.4.13) follows from Hadamard's inequality, since the diagonal elements of $\mathcal{I}(d)$ are equal to 1. Equality in (2.4.13) holds if and only if $\mathcal{I}(d) = \mathbf{X}'\mathbf{D}\mathbf{X}$ is diagonal. Thus $|\mathcal{I}(d)|$ attains its maximum, i.e. $d$ is D-optimum design, if and only if $\mathcal{I}(d) = \mathbf{I}_{m+1}$. Since $\mathrm{tr}[\mathcal{I}(d)] = \lambda_1 + \lambda_2 + \cdots + \lambda_{m+1} = m+1$, also the smallest eigenvalue of $\mathcal{I}(d)$ and $\phi_{-1}[\mathcal{I}(d)]$ attain their maximum value if and only if $\mathcal{I}(d) = \mathbf{I}_{m+1}$. As noted earlier, a design with $\mathcal{I}(d) = \mathbf{I}_{m+1}$ always exists, and hence only such designs are A- , D- , and E- optimal designs. This concludes the proof. □

**Remark 2.4.1** Though the complete factorial design assigning uniform weight $1/2^m$ to each of the $n = 2^m$ vertices of the $m$ dimensional cube $[-1,1]^m$ is optimal, its support size $2^m$ grows very quickly when the dimension $m$ increases.

For certain values of $m$ there exists a design $d^{(m+1)} = \{\mathbf{x}_1, \mathbf{x}_2, \dots, \mathbf{x}_{m+1}; p_1, p_2, \dots p_{m+1}\}$ with the minimum support size $m+1$ for which $\mathcal{I}(d^{m+1}) = \mathbf{I}_{m+1}$. An $(m+1) \times (m+1)$ matrix $\mathbf{X}$ with entries 1 and $-1$ is called a *Hadamard matrix* (denoted by $\mathbf{H}_{m+1}$) if $\mathbf{X}'\mathbf{X} = (m+1)\mathbf{I}_{m+1}$. Then the model matrix $\mathbf{X}$ is square and hence $|\mathbf{X}'\mathbf{D}\mathbf{X}| = |\mathbf{D}\mathbf{X}\mathbf{X}'|$. Now by Hadamard's inequality

$$\left|\mathcal{I}(d^{(m+1)})\right| = \lambda_1 \lambda_2 \cdots \lambda_{m+1} \leq \prod_{i=1}^{m+1} p_i(1 + \mathbf{x}_i'\mathbf{x}_i) \leq 1. \tag{2.4.14}$$

Equality holds if and only if the matrix $\mathbf{X}\mathbf{X}'$ is diagonal, that is if $1 + \mathbf{x}_i'\mathbf{x}_j = 0$ for all $i \neq j \leq m+1$, and $p_1 = p_2 = \cdots = p_{m+1} = \frac{1}{m+1}$.

If the model matrix $\mathbf{X}$ of an $(m+1)$-point design $d^{(m+1)}$ is an $(m+1) \times (m+1)$ Hadamard matrix $\mathbf{H}_{m+1}$, then $\mathcal{I}(d^{(m+1)}) = \mathbf{I}_{m+1}$. Note that the design $d^{(m+1)}$ with $\mathcal{I}(d^{(m+1)}) = \mathbf{I}_{m+1}$ is a uniform simplex design which assigns weight $1/(m+1)$ to each of the vertices $\mathbf{x}_1$, $\mathbf{x}_2$, ..., $\mathbf{x}_{m+1}$ of a regular simplex. The support points $\mathbf{x}_i \in \{\pm 1\}^m$, $i = 1, 2, \dots, m+1$ are also vertices of the $m$ dimensional cube $[-1,1]^m \subset \mathcal{T}_{\sqrt{m}}$. It is known that if a Hadamard matrix of order $k$ exists then $k = 2$ or $k = 4q$ for some positive integer $q$. Although it has not yet been shown that Hadamard matrices of order $4q$ exist for all $q \geq 1$, many infinite families of Hadamard matrices have been constructed. These include all values of $q$ which are of practical interest. A useful reference is Hedayat and Wallis (1978).

In conclusion, we note that optimal designs with both the support sizes $m+1$ and $2^m$ are available whenever $\mathbf{H}_{m+1}$ exists. It would be interesting to examine what other support sizes also provide optimal desgns.

In this context, an important result due to Caratheodory (Caratheodory's theorem, *see* Section 1.1) provides useful guidance. See also Pukelsheim (1993) and Silvey (1980). In the present set-up, we have a total of $m+1$ parameters. According to Caretheodory's theorem, any information matrix based on an arbitrary design with $n$ distinct support points can be realized by an alternative design with *at the most* $(m+2)(m+1)/2+1$ distinct support points. The following table serves to indicate situations where Caretheodory's theorem leads to a substantial reduction in the number of support points of a design.

**Table 2.1.** Support size of designs with information matrix $\mathcal{I}_{m+1}$.

| $m$ | $m+1$ | Factorial bound* | Hadamard bound** | Caratheodory upper bound |
|---|---|---|---|---|
| 3 | 4 | 8 | 4 | 11 |
| 4 | 5 | 16 | - | 16 |
| 5 | 6 | 32 | - | 22 |
| 6 | 7 | 64 | - | 29 |
| 7 | 8 | 128 | 8 | 37 |
| 8 | 9 | 256 | - | 46 |

*Based on Complete Factorial Experiment with levels $\pm 1$.

**Based on Hadamard matrices (with entries $\pm 1$).

For $m+1 = 0$ ($mod$ 4), Hadamard bound gives an optimal design with minimum support size and hence may be recommended unconditionally. In all other situations (except $m = 3$), Caratheodory's upper bound gives a substantial reduction of the support size as against factorial bound. However, the point to be noted is that such a design may *not* be easy to conceive and to actually construct. On the other hand, designs based on the factorial bound are very easy to construct and the only objection against them is their large support sizes. Still it seems that from a practitioner's point of view, the choice essentially lies in between designs based on the factorial bound and those based on the Hadamard bound (whenever the latter is available).

## References

de la Garza, A. (1954). Spacing of information in polynomial regression. *Annals of Mathematical Statistics* **25**, 123–130.

Elfving, G. (1952). Optimum allocation in linear regression theory. *Annals of Mathematical Statistics* **23**, 255–262.

Elfving, G. (1959). Design of linear experiments. *Probability and statistics. The Harald Cramér volume* (ed. by Ulf Grenander), 58–74. Wiley, New York.

Hedayat, A. S. and Wallis, W. D. (1978). Hadamard matrices and their applications. *Annals of Statistics* **6**, 1184–1238.

Horn, R. A. and Johnson, C. R. (1985). *Matrix analysis.* Cambridge: Cambridge University Press.

Kiefer, J. C. (1959). Optimum experimental designs (with discussion). *Journal of the Royal Statistical Socciety* Series B **21**, 272–319.

Liski, E. P. and Zaigraev, A. (2001). A stochastic cracterization of Loewner optimality design criterion in linear models. Appears in *Metrika.*

Pukelsheim, F. (1993). *Optimal design of experiments.* Wiley, New York.

Shah, K. R. and Sinha, Bikas K. (1989). Theory of optimal designs. *Lecture Notes in Statistics* **54**, Springer, New York.

Silvey, S. D. (1980). *Optimum design.* Chapman & Hall, London.

# 3

# Optimal Regression Designs in Asymmetric Domains

## Summary


**Features**

**Model(s):** Fixed coefficients regression (FCR), random coefficients regression (RCR) models (single factor linear, quadratic and cubic)
**Experimental domains:** [0, 1], [0, $h$] and ($h$, $H$]
**Optimality criteria:** Minimization of optimality functionals
**Major tools:** de la Garza (DLG) phenomenon and Loewner order domination (LOD) of information matrices for search reduction
**Optimality results:** Specific optimal designs under regression models for estimation, prediction and inverse prediction - all under continuous design theory
**Thrust:** Asymmetric experimental domains

Most of the results on optimum regression designs deal with the symmetric experimental domain $\mathcal{T} = [-1, 1]$. However, in general, optimality criteria do not enjoy invariance property with respect to the experimental domains. Keeping this in view, optimality problems in the specific asymmetric experimental domain of the form $\mathcal{T} = [0, 1]$ are discussed. As in Chapter 2, here also we have tried to exploit the DLG phenomenon to the extent possible and we have attained substantial reduction, in the LOD sense, in our search for optimal designs. Specific optimal designs have been derived for the problems of estimation of parameters (over the experimental domain $\mathcal{T} = [0, 1]$), prediction (over the domain $[h, H]$) and inverse prediction (over the domain $\mathcal{T} = [0, h]$). The results have been discussed in the framework of linear and quadratic regression under FCR and RCR models. Also some results for cubic regression under FCR model are reported. All the results are derived using continuous design theory.

## 3.1 Introduction

In this Chapter we discuss various interesting aspects of the famous *de la Garza phenomenon* (DLG phenomenon) in the set-up of linear and quadratic regression designs in the study of Loewner order domination (LOD) of information matrices. Emphasis is given on the exploitation of the DLG phenomenon and LOD in reducing the dimension of the problem of determining optimal regression designs by providing an *essentially complete class (ECC)*. The entire study is based on approximate design theory and the experimental domain is taken to be asymmetric such as $[0,1]$ or $[0,h]$.

The vast literature on optimal polynomial regression designs (involving a single non-stochastic regressor) centers around the specific symmetric experimental domain $\mathcal{T} = [-1,\ 1]$. It is not clear how far such an optimality study naturally extends itself to other experimental domains - especially to an asymmetric experimental domain of the form $\mathcal{T}_0 = [0,\ 1]$ - though it is very true that $-1 \le x \le 1$ if and only if $0 \le z \le 1$ where $z = (1+x)/2$. In this sense

$$\begin{aligned} E(Y_x) &= \beta_0 + \beta_1 x + \ldots + \beta_k x^k \\ &= \gamma_0 + \gamma_1 z + \ldots + \gamma_k z^k, \end{aligned}$$

where $\boldsymbol{\gamma} = \mathbf{L}\boldsymbol{\beta}$ for a suitably defined $\mathbf{L}$ derived through the transformation $z = (1+x)/2$. Note that $\mathbf{L}$ is of full rank and independent of $x$.

Next, starting with an $n$-point design $d_n(\mathbf{x},\mathbf{p})$, we have

$$\mathcal{I}_{\mathbf{x}}(\boldsymbol{\beta}) = \sum_{i=1}^{n} p_i \mathbf{f}(x_i)\mathbf{f}'(x_i),$$

where $\mathbf{f}(x) = (1, x, x^2, \ldots, x^k)'$ and $\mathbf{x} = (x_1, x_2, \ldots, x_n)$. Further, $d_n(\mathbf{x},\mathbf{p})$ is equivalent to $d_n^*(\mathbf{z},\mathbf{p})$ and

$$\mathcal{I}_{\mathbf{z}}(\boldsymbol{\gamma}) = \sum_{i=1}^{n} p_i \mathbf{f}(z_i)\mathbf{f}'(z_i),$$

where $\mathbf{z} = (z_1, z_2, \ldots, z_n)'$, $\mathbf{p} = (p_1, p_2, \ldots, p_n)'$ and $\boldsymbol{\gamma} = (\gamma_0, \gamma_1, \ldots, \gamma_k)'$.

It follows that $\mathcal{I}_{\mathbf{x}}(\boldsymbol{\beta}) = \mathbf{L}'\mathcal{I}_{\mathbf{z}}(\boldsymbol{\gamma})\mathbf{L}$. An immediate consequence of this relation is that both the DLG phenomenon and LOD carry over to $\mathcal{T}_0$ from $\mathcal{T}$ and vice verse. It is readily observed that the above correspondence also holds between $\mathcal{T} = [-1,\ 1]$ and any arbitrary experimental domain $[a,\ b]$ in the form of a non-degenerate finite interval. The pertinent question to be raised now is the following: Suppose that a criterion-specific optimal design $d(\mathbf{x},\ \mathbf{p})$ has been constructed with respect to $\mathcal{T} = [-1,\ 1]$. Does it naturally lend itself to the corresponding optimal design $d(\mathbf{z},\ \mathbf{p})$ over $\mathcal{T}_0 = [0,\ 1]$, obtained by the transformation from $x$ to $z$?

We believe that the only optimality criterion that enjoys this nice property is the D-optimality criterion (Recall the relation between $\mathcal{I}_{\mathbf{x}}(\boldsymbol{\beta})$ and

$\mathcal{I}_z(\gamma)$). Surprisingly, even for A- and E-optimality criteria, this transformation does *not* lead to an optimal design over [0, 1] from the corresponding design over [−1, 1]. This has prompted us to initiate a close and critical study for the quadratic and cubic regression over [0, 1]. It is interesting to note that in the literature there are instances of study of quadratic regression over [0, 1] (vide Atkinson and Donev 1992 and Pázman 1986). These results will be reported in proper places.

We organize this Chapter as follows. In this Section, we start with the specific framework of linear regression in order to place the DLG phenomenon in its proper perspective. Then in Section 3.2, we study quadratic and cubic regression models with and/or without the intercept term. In Section 3.3, we undertake a detailed study of optimal regression designs for parameter estimation in the framework of linear and quadratic regression with random regression coefficients. Next, in Section 3.4, we deal with the prediction problems. Finally, in Section 3.5, we discuss the problems of characterization of optimal regression designs for inverse prediction. All throughout, the emphasis is on asymmetric experimental domains.

Material in this Chapter is largely based on three papers by Liski *et al.* (1996,1997,1998) and a paper by Mandal *et al.* (2000). Two recent manuscripts by Luoma *et al.* (2001a, 2001b) are also cited for cubic regression.

Consider a first-degree model where we have $N$ uncorrelated responses

$$Y_{ij} = \beta_0 + \beta_1 x_i + e_{ij}, \tag{3.1.1}$$

$i = 1, 2, \ldots, n$ and $j = 1, 2, \ldots, N_i$ ($\sum_{i=1}^{n} N_i = N$) with expectations and variances given by

$$E(Y_{ij}) = \beta_0 + \beta_1 x_i \quad \text{and} \quad V(Y_{ij}) = \sigma^2 \tag{3.1.2}$$

respectively. We take the experimental domain to be $\mathcal{T} = [0, 1]$. Suppose for simplicity that $\sigma^2 = 1$.

Note that the exact design considered above is $d_{n:N}$ as in (1.2.4) and its continuous design theory analogue is given by (1.2.5) *viz.*,

$$d_n = (x_1, x_2, \ldots, x_n; p_1, p_2, \ldots, p_n). \tag{3.1.3}$$

In the exact theory sense, the dispersion matrix of the least squares estimates (LSE) of the parameters $\boldsymbol{\beta} = (\beta_0, \beta_1)'$ is given by

$$\mathbf{V}(\tilde{\boldsymbol{\beta}}) = \frac{1}{N} \begin{pmatrix} 1 & \sum_{i=1}^{n} \frac{N_i}{N} x_i \\ \sum_{i=1}^{n} \frac{N_i}{N} x_i & \sum_{i=1}^{n} \frac{N_i}{N} x_i^2 \end{pmatrix}^{-1}. \tag{3.1.4}$$

On the other hand, according to the continuous design theory, the information matrix (per observation) for the parameters is given by

$$\mathcal{I}(d) = \sum_{i=1}^{n} p_i \begin{pmatrix} 1 \\ x_i \end{pmatrix} \begin{pmatrix} 1 \\ x_i \end{pmatrix}' = \begin{pmatrix} 1 & \mu_1 \\ \mu_1 & \mu_2 \end{pmatrix}, \tag{3.1.5}$$

where $\mu_1 = \sum_{i=1}^{n} p_i x_i$ and $\mu_2 = \sum_{i=1}^{n} p_i x_i^2$.

Let $d_{2:p} = \{a, b;\ p, 1-p\}$ denote a 2-point design with weights $0 < p < 1$ and $1 - p = q$ at the points $a$ and $b$ respectively. Let $d_n$ be any $n$-point design (3.1.3) for the LSE of $\boldsymbol{\beta}$ in (3.1.1). Then the DLG phenomenon (de la Garza 1954) states that there exists a 2-point design $d_{2:p}$ such that $\min\{x_1, x_2, \ldots, x_n\} \le a < b \le \max\{x_1, x_2, \ldots, x_n\}$ and $\mathcal{I}(d_{2:p}) = \mathcal{I}(d_n)$. A quick proof follows.

Note first that

$$\mathcal{I}(d_{2:p}) = p\begin{pmatrix} 1 & a \\ a & a^2 \end{pmatrix} + q\begin{pmatrix} 1 & b \\ b & b^2 \end{pmatrix} = \begin{pmatrix} 1 & pa+qb \\ pa+qb & pa^2+qb^2 \end{pmatrix}, \tag{3.1.6}$$

where $q = 1 - p$. Choose $a = x_{\text{min}}$, for example, and find the values of $0 < p < 1$ and $b \le x_{\text{max}}$ such that $\mu_1 = px_{\text{min}} + (1-p)b$ and $\mu_2 = px_{\text{min}}^2 + (1-p)b^2$. Routine calculations give

$$b = \frac{\mu_2 - \mu_1 x_{\text{min}}}{\mu_1 - x_{\text{min}}}, \quad p = \frac{\mu_2 - \mu_1^2}{(\mu_1 - x_{\text{min}})(b - x_{\text{min}})}. \tag{3.1.7}$$

Thus the information matrices $\mathcal{I}(d_n)$ and $\mathcal{I}(d_{2:p})$ are equal. We say that the designs $d_n$ and $d_{2:p}$ are *information equivalent.* Since any comparison between designs will be based solely on their information matrices, we can confine our study of a line fit model (3.1.1) to the class of 2-point designs.

We now go ahead and compare designs in the class of 2-point designs with respect to the *Loewner ordering.* Recall Loewner domination concept as discussed in the beginning of Section 2.2. Theorem 3.1.1 shows that there exists a Loewner superior subclass among the 2-point designs.

**Theorem 3.1.1** *For any given $d_{2:p} = \{a, b;\ p, 1-p\}$ with $0 < a < b < 1$, there exists a 2-point design $d_{2:p}^* = \{0, 1;\ p^*, 1-p^*\}$ that dominates $d_{2:p}$, i.e.*

$$d_{2:p}^* = \{0, 1;\ p^*, 1-p^*\} \succ d_{2:p} = \{a, b;\ p, 1-p\}. \tag{3.1.8}$$

**Proof.** Utilization of (3.1.6) yields

$$\mathcal{I}(d_{2:p}^*) - \mathcal{I}(d_{2:p}) = \begin{pmatrix} 0 & 1-p^* - [pa + (1-p)b] \\ 1-p^* - [pa + (1-p)b] & 1-p^* - [pa^2 + (1-p)b^2] \end{pmatrix}. \tag{3.1.9}$$

Henceforth, we will abbreviate 2-point designs of the type $d_{2:p}$ simply as $d_p$. For any given $d_p$ with $0 < a < b < 1$ we can always choose $p^* = [1 - pa - (1-p)b]$ so that the nondiagonal elements in (3.1.9) become zero. Then clearly $1 - p^* - [pa^2 + (1-p)b^2] > 0$, and consequently $\mathcal{I}(d_p^*) - \mathcal{I}(d_p)$ is nonnegative definite. Thus there always exists $d_p^* \ne d_p$ such that $d_p^* \succ d_p$. □

Let us now suppose that there exists a 2-point design $d_w = \{0, 1;\ w, 1-w\}$ that dominates $d_p = \{0, 1;\ p, 1-p\}$ with respect to Loewner ordering, i.e. $d_w \succ d_p$. Then by (3.1.9), $\mathcal{I}(d_w) \ge \mathcal{I}(d_p)$ if and only if $w = p$. Given any 2-point design $d_p = \{0, 1;\ p, 1-p\}$, there exists no competing 2-point design $d_w = \{0, 1;\ w, 1-w\}$, $w \neq p$, which dominates $d_p$. Thus every 2-point design $d_p = \{0, 1;\ p, 1-p\}$ with $0 < p < 1$ is *admissible*. We also say that the information matrices of those designs are admissible. Correspondingly, every design $d_s = \{a, b;\ s, 1-s\}$ with support points $0 < a < b < 1$ is *inadmissible* and there exists by Theorem 3.1.1 an admissible design which dominates it. This means that the admissible designs form a *complete class*. For definitions and more extensive discussion on these concepts see Pukelsheim (1993, Chapter 10).

The above results have far-reaching implications, since all comparisons of designs are based on the information matrices, and any reasonable *design optimality criterion* is *isotonic* with respect to the Loewner ordering.

Thus the class of 2-point designs $d_p = \{0, 1; p, 1-p\}$, $0 < p < 1$, is an essentially complete class of designs for the model (3.1.1). In Section 3.2, we deal with the quadratic regression models and develop the DLG phenomenon. In the process we supplement some results already available in the literature (vide Pázman 1986, Atkinson and Donev 1992).

**Remark 3.1.1** Once more recall Kiefer's result, as explained in Remark 2.3.1, for discarding a vast subclass of competing designs in a polynomial regression set-up. He establsihed that in case of a polynomial regression over $[-1,\ 1]$, any design which does *not* include both the extreme points $1$ and $-1$ must be inadmissible and hence left out. This result, when translated for $[0,\ 1]$ states that effectively, both 0 and 1 must be included among the support points of admissible designs. In Theorem 3.1.1, we have only given a direct verification of this result in case of linear regression. For quadratic regression again, an explicit verification is presented in the next section.

## 3.2 de la Garza Phenomenon in Quadratic and Cubic Regression

We will deal with fixed coefficient quadratic and cubic regression models with homoscedastic errors and study various aspects of optimal designs for estimation of underlying parameters. We start with a quadratic model of the form

$$Y_x = \beta_0 + \beta_1 x + \beta_2 x^2 + e_x \qquad (3.2.1)$$

and the experimental domain is again taken to be $\mathcal{T} = [0, 1]$. Then according to the DLG Phenomenon, given an arbitrary $n$ ($> 3$) point design, we can choose a 3-point design such that the information matrices of the parameter vector $\boldsymbol{\beta} = (\beta_0, \beta_1, \beta_2)'$ for the two designs are identical. Thus, we can confine to the class of three-point designs only.

### 3.2.1 Quadratic Regression with Full Set of Parameters

Let $d_3 = \{x_1, x_2, x_3; p_1, p_2, p_3\}$ be any 3-point design such that $0 < x_1 < x_2 < x_3 < 1$. Then the elements $\mu_{jk}$ of the information matrix $\mathcal{I}(d_3)$ are

$$\mu_{jk} = \sum_{i=1}^{3} p_i x_i^{j+k-2}; \quad j, k = 1, 2, 3. \tag{3.2.2}$$

We are able to establish a non-trivial dominance result.

**Theorem 3.2.1** *Given $d_3$ as above, there exists a 3-point design $\tilde{d}_3 = \{0, \tilde{x}, 1;\ \tilde{p}_1, \tilde{p}_2, \tilde{p}_3\}$ with $0 < \tilde{x} < 1$ such that $\tilde{d}_3 \succ d_3$.*

**Proof.** We will abbreviate both $\tilde{d}_3$ and $d_3$ simply as $\tilde{d}$ and $d$ respectively. The entries $\tilde{\mu}_{jk}$ of $\mathcal{I}(\tilde{d})$ are of the form

$$\tilde{\mu}_{jk} = \tilde{p}_2 \tilde{x}^{j+k-2} + \tilde{p}_3; \quad j, k = 1, 2, 3. \tag{3.2.3}$$

Note that $\mu_{22} = \mu_{13}$ and $\tilde{\mu}_{22} = \tilde{\mu}_{13}$. For a given design $d$, we have the moments $\mu_{jk}$, $j$, $k = 1$, 2, 3. We equate the corresponding entries of $\mathcal{I}(d)$ and $\mathcal{I}(\tilde{d})$, except $\mu_{33}$ and $\tilde{\mu}_{33}$. Then we have the equations

$$\begin{aligned} \tilde{p}_2 \tilde{x} + \tilde{p}_3 &= \mu_{12} \\ \tilde{p}_2 \tilde{x}^2 + \tilde{p}_3 &= \mu_{13} (= \mu_{22}) \\ \tilde{p}_2 \tilde{x}^3 + \tilde{p}_3 &= \mu_{23}. \end{aligned} \tag{3.2.4}$$

It follows that

$$\tilde{x} = \frac{\mu_{13} - \mu_{23}}{\mu_{12} - \mu_{13}}. \tag{3.2.5}$$

Now solving for $\tilde{p}_2$ and $\tilde{p}_3$ from (3.2.4) we obtain

$$\tilde{p}_2 = (\mu_{12} - \mu_{13})/(\tilde{x} - \tilde{x}^2) \ \text{ and } \ \tilde{p}_3 = \mu_{12} - \tilde{p}_2 \tilde{x}. \tag{3.2.6}$$

It is readily verified that $0 < \tilde{x} < 1$. It now remains to establish that $0 < \tilde{p}_i < \tilde{p}_2 + \tilde{p}_3 < 1$, for $i = 2, 3$. We do this in the Appendix A3.2.1. Therefore, given a design $d$, we can choose a design $\tilde{d} = \{0, \tilde{x}, 1;\ \tilde{p}_1, \tilde{p}_2, \tilde{p}_3\}$ such that $\tilde{x}$, $\tilde{p}_2$, $\tilde{p}_3$ ($\tilde{p}_1 = 1 - \tilde{p}_2 - \tilde{p}_3$) is a feasible solution of (3.2.4). Then all entries of $\mathcal{I}(\tilde{d}) - \mathcal{I}(d_3)$ are zero, except

$$\begin{aligned} \tilde{\mu}_{33} - \mu_{33} &= \tilde{p}_2 \tilde{x}^4 + \tilde{p}_3 - \mu_{33} \\ &= \tilde{p}_2 (\tilde{x}^4 - \tilde{x}) + (\mu_{12} - \mu_{33}). \end{aligned}$$

Note that $\tilde{x}^4 \leq \tilde{x}$ though $\mu_{12} \geq \mu_{33}$. Therefore, we need to argue that the above quantity is strictly positive. This argument is carried out in the Appendix A3.2.2. Thus, finally, it follows that $\tilde{d} \succ d$ with respect to Loewner ordering and $\tilde{d} \neq d$. This concludes the proof. □

**Remark 3.2.1** The above dominance result is extremely helpful in the actual characterization of specific optimal designs. This is done below. However, the point to be noted is that within the class of 3-point designs, further Loewner order domination is *not* generally feasible so that no *best* design exists. In spite of this limitation, it would be desirable to examine the nature of complete classes of designs for general polynomial regression models. However, the results may not be as useful as in the case of linear or quadratic regression. Following Kiefer (1959), it is evident that at best we may suggest that such designs will include the extreme points of the experimental domain. Thus, even for a third degree polynomial regression model, it may not be enough to know just that! Still five parameters (two intermediate points and weights at some three out of four points) remain to be determined. This has recently been studied in Luoma *et al.* (2001a) and will be reported later in Subsection 3.2.3.

**Specific Optimality Results for the full set of parameters**

As background information, we may note that

(a) $\mathcal{I} = \begin{pmatrix} 1 & 1 & 1 \\ 0 & x & 1 \\ 0 & x^2 & 1 \end{pmatrix} \begin{pmatrix} p_1 & 0 & 0 \\ 0 & p_2 & 0 \\ 0 & 0 & p_3 \end{pmatrix} \begin{pmatrix} 1 & 0 & 0 \\ 1 & x & x^2 \\ 1 & 1 & 1 \end{pmatrix}$

(b) $|\mathcal{I}| = p_1 p_2 p_3 x^2 (1-x)^2$

(c) $\mathcal{I}^{-1} = \frac{1}{x^2(1-x)^2} \begin{pmatrix} x(1-x) & 0 & 0 \\ -(1-x^2) & 1 & -x^2 \\ 1-x & -1 & x \end{pmatrix} \begin{pmatrix} p_1^{-1} & 0 & 0 \\ 0 & p_2^{-1} & 0 \\ 0 & 0 & p_3^{-1} \end{pmatrix} \times \begin{pmatrix} x(1-x) & -(1-x^2) & 1-x \\ 0 & 1 & -1 \\ 0 & -x^2 & x \end{pmatrix}$

Though D-optimal designs can be obtained by transforming the D-optimal designs over $[-1,\ 1]$, we present them directly for the sake of completeness in several cases below.

(i) *A-optimal design*: $\{0,\ 0.49976,\ 1.0;\ 0.3217,\ 0.4862,\ 0.1921\}$[1]

(ii) *D-optimal design*: $\{0,\ 0.5,\ 1.0;\ 1/3,\ 1/3,\ 1/3\}$

The details are now outlined below.

We denote an optimal design by the generic notation $[0,\ \tilde{x},\ 1;\ \tilde{p_1},\ \tilde{p_2},\ \tilde{p_3}]$ with $\tilde{p_1} + \tilde{p_2} + \tilde{p_3} = 1$. Then the expressions to be minimized are given by:

**A-optimality**:

$$\frac{A_1}{\tilde{p_1}} + \frac{A_2}{\tilde{p_2}} + \frac{A_3}{\tilde{p_3}}, \tag{3.2.7}$$

[1]Professor Olaf Kraft communicated an alternative derivation of this solution to one of the authors. The derivation uses Kiefer's equivalence Theorem.

where

$$\begin{aligned} A_1 &= 2(1-\tilde{x}-\tilde{x}^3+\tilde{x}^4)/\tilde{x}^2(1-\tilde{x})^2 \\ A_2 &= 2/\tilde{x}^2(1-\tilde{x})^2 \\ A_3 &= (1\ +\ \tilde{x}^2)/(1-\tilde{x})^2 \end{aligned} \tag{3.2.8}$$

whence $\tilde{p}_{iopt} = \sqrt{A_i}/(\sqrt{A_1}+\sqrt{A_2}+\sqrt{A_3})$; $i = 1, 2, 3$. We have then determined $\tilde{x}_{opt}$ by minimizing

$$[\sqrt{2}(1-\tilde{x}-\tilde{x}^3+\tilde{x}^4)+\sqrt{2}+\tilde{x}\sqrt{(1+\tilde{x}^2)}]/\tilde{x}(1-\tilde{x})]$$

and hence $\tilde{p}_{iopt}$ for $i = 1, 2, 3$.

**D-optimality**:

$$[\tilde{p}_1\tilde{p}_2\tilde{p}_3\tilde{x}^2(1-\tilde{x})^2]^{-1}. \tag{3.2.9}$$

It is readily seen that the design given above is $D$-optimal.

Pázman (1986, Chapter 1, Example 3 ) describes the nature of optimal design for efficient estimation of $\beta_0-\beta_1+\beta_2 = (1,\ -1,\ 1)'\boldsymbol{\beta}$. In our set-up, it would mean that we have to choose $\tilde{x}$, $\tilde{p}_1$, $\tilde{p}_2$, $\tilde{p}_3$ so that

$$f = (1,\ -1,\ 1)'\mathcal{I}^{-1}(1,\ -1,\ 1) \tag{3.2.10}$$

is minimized.

Routine calculations yield that

$$f = \left[\frac{4(1-\tilde{x}^2)^2}{\tilde{p}_1}+\frac{4}{\tilde{p}_2}+\frac{\tilde{x}^2(1+\tilde{x})^2}{\tilde{p}_3}\right]/\tilde{x}^2(1-\tilde{x})^2,$$

whence

$$\tilde{p}_{1opt} = \frac{2(1-\tilde{x}^2)}{4+\tilde{x}-\tilde{x}^2},\ \tilde{p}_{2opt} = \frac{2}{4+\tilde{x}-\tilde{x}^2},\ \tilde{p}_{3opt} = \frac{\tilde{x}(1+\tilde{x})}{4+\tilde{x}-\tilde{x}^2}.$$

This gives

$$f = \frac{16}{\tilde{x}^2(1-\tilde{x})^2}+\frac{8}{\tilde{x}(1-\tilde{x})}+1.$$

Thus, finally, we obtain: $\tilde{x}_{opt} = 0.5$ and $\tilde{p}_{1opt} = 0.35$, $\tilde{p}_{2opt} = 0.47$ and $\tilde{p}_{3opt} = 0.18$. These results are in agreement with Pázman (1986).

Unlike A- and D-optimality criteria, the other two viz., E- and MV-optimality criteria do *not* exhibit meaningful statistical interpretation when the parameter vector includes the intercept term. Keeping this in view, we discuss about the MV-optimality criterion in the context of subset estimation of parameters, excluding the intercept term.

**Specific Optimality Results for the Subset of Parameters Excluding the Intercept Term**

For ready reference, we note that $\mathcal{I}^{-1}$ (displayed in (c) above) involves linear functions of $p_1^{-1}, p_2^{-1}$ and $p_3^{-1}$ along the principal diagonal (and elsewhere as well). Further, $|\mathcal{I}_{22.1}| = |\mathcal{I}|$.

(i) *A-optimal design:* $\{0,\ 0.50198,\ 1.0;\ 0.328406,\ 0.480683,\ 0.190911\}$

(ii) *D-optimal design:* $\{0,\ 0.5,\ 1.0; 1/3,\ 1/3\ , 1/3\}$

(iii) *MV-optimal design:* $\{0,\ 0.89995,\ 1.0;\ 0.072829,\ 0.497974,\ 0.429197\}$

Adopting the same notations as before, the expressions to be minimized are given by:

**A-optimality:**

$$\frac{A_1}{\tilde{p_1}} + \frac{A_2}{\tilde{p_2}} + \frac{A_3}{\tilde{p_3}}, \tag{3.2.11}$$

where

$$\begin{aligned} A_1 &= (2 - 2\tilde{x} - \tilde{x}^2 + \tilde{x}^4)/\tilde{x}^2(1-\tilde{x})^2 \\ A_2 &= 2/\tilde{x}^2(1-\tilde{x})^2 \\ A_3 &= (1 + \tilde{x}^2)/(1-\tilde{x})^2. \end{aligned} \tag{3.2.12}$$

The rest is along the same lines as before.

For D-optimality, the same optimality result continues to hold.

**MV-optimality:**

$$[\tilde{x}^2\tilde{p_2}(1-\tilde{p_2}) + \tilde{p_3}(1-\tilde{p_3} - 2\tilde{x}\tilde{p_2}\tilde{p_3}]/\tilde{p_1}\tilde{p_2}\tilde{p_3}\tilde{x}^2(1-\tilde{x})^2 \tag{3.2.13}$$

subject to

$$2\tilde{p_3} = \tilde{x}(1+\tilde{x})(1-\tilde{p_2}) \tag{3.2.14}$$

which simplifies to

$$\left[\frac{A}{(1-\tilde{p_2})} + \frac{1}{\tilde{p_2}}\right] /[\tilde{x}^2(1-\tilde{x})^2] \tag{3.2.15}$$

with

$$A = [2(1-\tilde{x}^3)/(2-\tilde{x}(1+\tilde{x}))] - [2\tilde{x}^2/(1+\tilde{x})]. \tag{3.2.16}$$

This leads to the minimization of

$$\tilde{x}^{-1}(1-\tilde{x})^{-1}[1 + \sqrt{\frac{2(1+\tilde{x}-2\tilde{x}^2)}{(1+\tilde{x})(2-\tilde{x}-\tilde{x}^2)}}]. \tag{3.2.17}$$

It is well-known that in order that a design is MV-optimal, the variance expressions for the estimators of the two parameters are to coincide and their common value is to be the least. Condition (3.2.14) referred to above corresponds to that of identity of the two variance expressions. The expression to be minimized in (3.2.13) is either of the two variance expressions. The rest is routine computation.

### 3.2.2 Quadratic Regression without the Intercept Term

Once again we want to look at the case of quadratic regression but with the intercept term missing, and wish to examine the implications of the DLG phenomenon in such cases. An optimality study along this direction is found in Atkinson and Donev (1992). See also Pázman (1986, Chapter 1, Example 3).

We will assume the experimental domain $\mathcal{T} = [0, 1]$. Since in $\mathcal{T}$, $x$ and $x^2$ are one to one functions of each other, we may note in passing that in the case where only the linear coefficient term is missing, we are essentially in the first degree model set-up after a change of regressor values from $x^2$ to $x$.

However, for the model

$$Y_x = \beta_1 x + \beta_2 x^2 + e_x \tag{3.2.18}$$

it would be interesting to examine the validity of the DLG phenomenon. In (3.2.18), $\beta_1$ is an indicator of the slope of the response function while $\beta_2$ measures its curvature. The response function is assumed to pass through the origin. See Pázman (1986) and Atkinson and Donev (1992) for practical examples. In this Monograph, we will consider the homoscedastic situation only.

To start with, let us suppose that there is a given $n$-point design having a support set with elements in $\mathcal{T}^* = (0, 1]$. Note that in the absence of $\beta_0$ (the intercept term), observations at the extreme point $x = 0$ are non-informative for the other parameters. The relevant moments for our study are: $\mu_2$, $\mu_3$ and $\mu_4$. Since there are two parameters, it would be interesting to examine if there is a 2-point design having the same set of moments as specified above. Suppose the 2-point design to be searched is denoted by $d = \{a, b; p, q\}$.

This would mean that we should be able to solve for $a$, $b$ and $p$ ($q = 1-p$) such that

$$\begin{aligned} &0 < a < b \le 1,\ 0 < p < 1 \text{ and } a^2p + b^2q = \mu_2, \\ &a^3p + b^3q = \mu_3 \text{ and } a^4p + b^4q = \mu_4. \end{aligned} \tag{3.2.19}$$

We will skip this aspect and, instead, pose and resolve two more pertinent questions:

Q.1 Can any $n$-point design (for $n \ge 3$) be Loewner order dominated by a 2-point design?

Q.2 Can any design $\{a, b; p, q\}$ be improved in the Loewner ordering sense?

The answer to Q.1 is definitely affirmative as we shall shortly see. Further, an affirmative answer to Q.2 is possible whenever $b < 1$ but only in the *weak* Loewner order domination sense to be explained below.

In case of the quadratic regression model (with all the parameters, including the intercept term), it was demonstrated earlier in Theorem 3.2.1

that Loewner order domination of any $n$-point ($n > 3$) design is possible by a 3-point design with two support points at the extremes i.e., at 0 and 1 and the third in the interior. It was also seen that in general, no further Loewner order domination is possible among 3-point designs. Here, however, for the model (3.2.18) and with reference to Q.2, we establish a result which provides a simplified approach in the search for specific optimal designs.

Whereas Loewner order domination of $\mathcal{I}^*$ over $\mathcal{I}$ means that $\mathcal{I}^* - \mathcal{I}$ is a nonnegative definite (nnd) matrix, *weak* Loewner order domination of order $l$ ($< k$), to be denoted by $\mathcal{I} \leq^{\omega} \mathcal{I}^*$, will refer to $\mathcal{I}_{11.2} \leq \mathcal{I}^*_{11.2}$, both of the latter matrices being of order $l \times l$. Here $\mathcal{I}_{11.2}$ refers to the information matrix for any subset of $l$ components of the parameter vector of $k$ components. The dominance property here refers to *all* possible principal submatrices of order $l \times l$ of $\mathcal{I}$ and $\mathcal{I}^*$, respectively.

We now take up the issues in Q.1 and Q.2 one by one. Note that in a standard quadratic regression model, a design with $n$ ($> 3$) points is information equivalent to a suitably chosen 3-point design. This is the essence of the DLG phenomenon. Further, we have noted in Theorem 3.2.1 that a 3-point design of the type $d = \{a, b, c; p, q, r\}$ with $0 < a < b < c < 1$ can be Loewner order dominated by another design of the type $d^* = [0, \lambda, 1; p^*, q^*, r^*]$. The proof was by construction and we ended up with the choice of $\lambda$, $p^*$, $q^*$ and $r^*$ in a way that the elements at the positions: (1,1), (1,2), (1,3) (= (2,2)) and (2,3) of $\mathcal{I}_d$ and $\mathcal{I}_{d^*}$ are the same while in the position (3,3), there is a strict inequality.

For the same choice of $d^*$ as against $d$, it thus follows that with reference to the model in (3.2.18), Loewner order domination of $d^*$ over $d$ continues to hold. Let us now write for the model (3.2.18) and for the design $d^* = \{0, \lambda, 1;\ p^*, q^*, r^*\}$ the information matrix

$$\mathcal{I}_{d^*} = \begin{pmatrix} A^* & B^* \\ & C^* \end{pmatrix} \tag{3.2.20}$$

and change $d^*$ to $d^{**} = \{\lambda, 1; q^*, p^* + r^*\}$. This yields for $d^{**}$ the information matrix

$$\mathcal{I}_{d^{**}} = \begin{pmatrix} A^* + p^* & B^* + p^* \\ & C^* + p^* \end{pmatrix}. \tag{3.2.21}$$

It is readily verified that $d^{**}$ Loewner order dominates over $d^*$ and, hence, over $d$.

To resolve Q.2, it is enough to change the design $\{a, b; p, q\}$ to $\{a/b, 1; p, q\}$. Note that

$$\mathcal{I}_{a,b;\ p,q} = p(a,\ a^2)'(a,\ a^2) \ +\ q(b,\ b^2)'(b,\ b^2)$$

while

$$\mathcal{I}_{a/b,1;\ p,q} = p(a/b,\ a^2/b^2)'(a/b,\ a^2/b^2) \ +\ q(1,\ 1)'(1,\ 1),$$

and write for simplicity $\mathcal{I}_{a,b;\ p,q} = \mathcal{I}$ and $\mathcal{I}_{a/b,1;\ p,q} = \mathcal{I}^*$. Hence, it follows that whenever $b < 1$,

$$\mathcal{I}_{11} = b^2\mathcal{I}^*_{11} < \mathcal{I}^*_{11},\quad \mathcal{I}_{22} = b^4\mathcal{I}^*_{22} < \mathcal{I}^*_{22};$$

$$\mathcal{I}_{11.2} = b^2\mathcal{I}^*_{11.2} < \mathcal{I}^*_{11.2},\quad \mathcal{I}_{22.1} = b^4\mathcal{I}^*_{22.1} < \mathcal{I}^*_{22.1}.$$

Consequently, the 2-point design $\{a/b, 1; p, q\}$ is better than the design $\{a, b; p, q\}$ with respect to weak Loewner order domination of order 1 and, hence, for A-, D- and MV-optimality criteria as well. It is not clear if this also holds for the E-optimality criterion. This latter design does not, however, improve over the former in the usual Loewner order domination sense. The revealing interpretation of the above is as follows: The class of 2-point designs including the extreme point 1 is *essentially complete but* there is no unique best 2-point design. However, in view of the above, the search for specific optimal designs is now drastically simplified. Before describing the explicit nature of optimal designs, the following remark is pertinent.

**Remark 3.2.2** In the above, we have employed a "scaling factor" to change the support points $a$ and $b$ to $a/b$ and 1. Another way out would be to use "location shift", thereby changing to $a + (1 - b)$ and 1. It can be easily verified that in this case keeping the same weights $p$ and $q$ located at the two positions respectively, the "shifted" design produces a lower value of the generalized variance than the former (i.e. the latter is "D-better" than the former). However, in general, we cannot even claim weak Loewner order dominance of the shifted designs.

Specific optimal designs are now derived below. For $E$-optimality as well, we confine to the class of competing designs including 1 in their support. The derivations are explained afterwards.

(i) *A-optimal design*: $\{0.419, 1.0; 0.757, 0.243\}$

(ii) *D-optimal design*: $\{0.5, 1.0; 0.5, 0.5\}$

(iii) *E-optimal design*: $\{0.414, 1.0; 0.767, 0.233\}$

(iv) *MV-optimal design*: $\{(\sqrt{2}-1), 1.0; \frac{1}{\sqrt{2}}, (1-\frac{1}{\sqrt{2}})\}$

We denote an optimal design by the generic notation $[c, 1;\ p, q]$ with $p+q = 1$. Note that the information matrix corresponding to $d = [c, 1;\ p, q]$ is

$$\mathcal{I}(d) = \begin{pmatrix} pc^2 + q & pc^3 + q \\ - & pc^4 + q \end{pmatrix}. \tag{3.2.22}$$

Then the expressions to be optimized are given by:

**A-optimality**:

$$\textit{Minimize}\ [c^2p + q + c^4p + q]/[pqc^2(1-c)^2] \tag{3.2.23}$$

which simplifies to

$$[(1+c^2)/(1-c)^2/q] + [2/c^2(1-c)^2]/p$$

and this is of the form: $A/q + B/p$ whence $p_{opt} = \sqrt{B}/(\sqrt{A}+\sqrt{B})$. We have then determined $c_{opt}$ by minimizing $[c\sqrt{(1+c^2)} + \sqrt{2}]/c(1-c)$ and hence determined $p_{opt}$.

**D-optimality**:

$$\textit{Minimize } [pqc^2(1-c^2)]^{-1}. \tag{3.2.24}$$

It is readily seen that the design given above is D-optimal. This result is *not* derivable from the corresponding result over $[-1,\ 1]$ by transforming to $[0,\ 1]$.

For E-optimality, we have to maximize the smaller of the two eigenvalues of the information matrix. Denoting by $\rho_1$ and $\rho_2$ the eigenvalues of $\mathcal{I}(d)$, we have

$$\rho_1 + \rho_2 = pc^2(1+c^2) + 2q, \quad \rho_1\rho_2 = pqc^2(1-c)^2$$

whence

$$(\rho_1 - \rho_2)^2 = 4q^2 + p^2c^4(1+c^2)^2 + 8pqc^3 = Q^2,$$

say. Therefore, the smaller of the two eigenvalues is $[pc^2(1+c^2)+2q-Q]/2 = \phi(p,c)$, *say*. Hence for

**E-optimality**:

$$\textit{Maximize } [pc^2(1+c^2) + 2q - Q]. \tag{3.2.25}$$

In order to maximize $\phi(p,c)$, we differentiate it with respect to $p$ and solve for $p$, for a given $c$. Then we maximize the resulting expression for variation in $c$, $0 < c < 1$. The equation in terms of $p$ is given by

$$c^2(1+c^2) - 2 = [2pc^4(1+c^2)^2 - 8q + 8c^3(q-p)]/2Q,$$

which simplifies to

$$4Q[c^2(1+c^2) - 2]^2 = [2pc^4(1+c^2)^2 - 8q + 8c^3(q-p)]^2$$

and further to a quadratic equation in $p$. The rest of the computation is routine.

**MV-optimality**:

$$\textit{Minimize } [c^2p + q]/[pqc^2(1-c)^2] \tag{3.2.26}$$

which simplifies to

$$[c^2/q + 1/p]/[c^2(1-c)^2] \tag{3.2.27}$$

whence

$$p_{opt} = 1/(1+c). \tag{3.2.28}$$

This leads to the minimization of $[1/c + 2/(1-c)]$. This gives $c_{opt} = \sqrt{2}-1$ and $p_{opt} = \frac{1}{\sqrt{2}}$.

Atkinson and Donev (1992) have suggested the nature of A-optimal design for this problem. It turns out that the above derivation is neat and also describes other optimal designs in a nice fashion. Pázman (1986) essentially derived the form of an optimal design for efficient estimation of $\beta_2$ in (3.2.18). For this problem, we need to maximize $\mathcal{I}_{22.11}$ and choose $c_{opt}$ and $p_{opt}$ accordingly. It turns out that $p_{opt} = \frac{1}{(1+c_{opt})}$ while $c_{opt} = \frac{1}{(1+\sqrt{2})}$. Thus, about 71 per cent of the observations are to be placed at 0.414 in the scale of $[0, 1]$ while the rest are to be placed at 1.

### 3.2.3 Optimal Designs in a Cubic Regression Model

In this subsection, following Luoma *et al.* (2001a), we will make a brief presentation of A-optimal designs under a cubic regression model when the experimental domain is $[0,\ 1]$ as against the symmetric domain $[-1,\ 1]$, as is normally studied (vide Pukelsheim 1993). Note that D-optimal designs under an asymmetric experimental domain can be obtained from the corresponding optimal designs under the symmetric experimental domain by the usual transformation. We will not, however, deal with E- or MV-optimality in the case of cubic regression.

We will denote by $\boldsymbol{\beta} = (\beta_0,\ \beta_1,\ \beta_2,\ \beta_3)'$ the parameter vector in the model which is an extension of (3.2.1).

**A-optimality: Estimation of all parameters**

Clearly, taking the clue from Kiefer (1959) (vide also Pukelsheim 1993), we know that all admissible designs take into account the extreme points 0 and 1 and we need two more intermediate points. We denote an optimal design by the generic notation $d_4(0,\ a,\ b,\ 1;\ p,\ q,\ r,\ s)$ with $p + q + r + s = 1$.

Also let

$$\boldsymbol{\Delta} = \begin{pmatrix} 1 & 1 & 1 & 1 \\ 0 & a & b & 1 \\ 0 & a^2 & b^2 & 1 \\ 0 & a^3 & b^3 & 1 \end{pmatrix}. \tag{3.2.29}$$

and denote by $\mathbf{D} = (d_{ij})$ the inverse of $\boldsymbol{\Delta}$. Then the expression to be minimized is given by:

$$\frac{A_1}{p} + \frac{A_2}{q} + \frac{A_3}{r} + \frac{A_4}{s} \tag{3.2.30}$$

where

$$\begin{aligned} A_1 &= d_{11}^2 + d_{12}^2 + d_{13}^2 + d_{14}^2 \\ A_2 &= d_{21}^2 + d_{22}^2 + d_{23}^2 + d_{24}^2 \\ A_3 &= d_{31}^2 + d_{32}^2 + d_{33}^2 + d_{34}^2 \\ A_4 &= d_{41}^2 + d_{42}^2 + d_{43}^2 + d_{44}^2 \end{aligned}$$

whence

$$p_{opt} = \sqrt{A_1}/(\sqrt{A_1} + \sqrt{A_2} + \sqrt{A_3} + \sqrt{A_4})$$

and similarly for the others. We have then determined $a_{opt}$ and $b_{opt}$ by minimizing the expression

$$\sqrt{A_1} + \sqrt{A_2} + \sqrt{A_3} + \sqrt{A_4}$$

and hence $p_{opt}$ etc may be easily computed.

*A-Optimal Design*
{0, 0.2516, 0.7479, 1.0; 0.2198, 0.3746, 0.2824, 0.1232}.

**A-optimality: Estimation of all parameters except the intercept term.**
We start with the same $d_4$ as before and note that the expression to be minimized is given by

$$\frac{A_1^*}{p} + \frac{A_2^*}{q} + \frac{A_3^*}{r} + \frac{A_4^*}{s} \tag{3.2.31}$$

where

$$\begin{aligned} A_1^* &= d_{12}^2 + d_{13}^2 + d_{14}^2 \\ A_2^* &= d_{22}^2 + d_{23}^2 + d_{24}^2 \\ A_3^* &= d_{32}^2 + d_{33}^2 + d_{34}^2 \\ A_4^* &= d_{42}^2 + d_{43}^2 + d_{44}^2 \end{aligned}$$

whence

$$p_{opt} = \sqrt{A_1^*}/(\sqrt{A_1^*} + \sqrt{A_2^*} + \sqrt{A_3})^* + \sqrt{A_4^*}$$

and similarly for the others. We have then determined $a_{opt}$ and $b_{opt}$ by minimizing

$$\sqrt{A_1^*} + \sqrt{A_2^*} + \sqrt{A_3^*} + \sqrt{A_4^*}$$

and hence $p_{opt}$ etc may be easily computed.

*A-Optimal Design*
{0, 0.2517, 0.7479, 1.0; 0.2193, 0.3748, 0.2826, 0.1233}.

**Remark 3.2.3** It is interesting to note that there is little variation in the nature of the above two optimal designs - one for estimation of all parameters and the other for all without the intercept term. We may note in passing that the D-optimal designs are, however, *exactly the same* for the above two problems and are easily derivable from the corresponding results over the interval $[-1,\ 1]$. The solution is

{0, 0.2765, 0.7235, 1.0; 0.25, 0.25, 0.25, 0.25}

(vide Pukelsheim 1993).

We do not discuss other variations of the model where the intercept term is absent. We refer to Luoma *et al.* (2001a) for some results is this area.

In the rest of this Chapter, we deal with first-degree and quadratic regression models with random coefficients. Again we will skip altogether discussion of corresponding results for a cubic regression with random coefficients. We refer to Luoma *et al.* (2001b) for some initial results in this direction.

## 3.3 Optimal Designs for Parameter Estimation in RCR Models

### 3.3.1 First-Degree RCR Model

. In view of Theorem 3.1.1, the search for criterion-based optimal designs has been reduced to the choice of $p$ in the class of 2-point designs $d_p = \{0, 1; p, 1-p\}$, where $p$ and $1-p$ are the weights at the design points 0 and 1 respectively. It turns out that such a choice will, in general, depend on the criterion as well as on the (ratio of the) variance components, i.e., the $\delta$ parameters. The case where we have only one random component and the other fixed, follows from the general case in a obvious way. Instead, we now demonstrate the nature of designs with specific optimality properties.

For subsequent convenience, we display the dispersion matrix of the least squares estimates of $\boldsymbol{\beta} = (\beta_0,\ \beta_1)'$ based on the above 2-point design as follows:

$$\mathbf{V}(\hat{\boldsymbol{\beta}}) = \begin{pmatrix} \delta_0 + \frac{1}{p} & -\frac{1}{p} \\ -\frac{1}{p} & \delta_1 + \frac{1}{p(1-p)} \end{pmatrix}. \tag{3.3.1}$$

In the following, we present the optimum values of $p$ for different optimality criteria.

**D-optimal Design**

We have to minimize the value of

$$|\mathbf{V}(\hat{\boldsymbol{\beta}})| = \delta_0\delta_1 + \frac{\delta_0}{p(1-p)} + \frac{\delta_1}{p} + \frac{1}{pq},$$

which yields

$$p_{opt} = \frac{\sqrt{\delta_0 + \delta_1 + 1}}{\sqrt{\delta_0 + 1} + \sqrt{\delta_0 + \delta_1 + 1}}. \tag{3.3.2}$$

**A-optimal Design**

A-optimal design minimizes

$$\mathrm{tr}[\mathbf{V}(\hat{\boldsymbol{\beta}})] = \delta_0 + \delta_1 + \left(\frac{1}{p} + \frac{1}{p(1-p)}\right)$$

and the minimum is attained when

$$p_{opt} = 2 - \sqrt{2}. \tag{3.3.3}$$

Thus A-optimal design is independent of $\delta_0$ and $\delta_1$. This also follows directly from Theorem 2.2.2.

**E-optimal Design**

A design is E-optimal if it minimizes the larger of the two eigenvalues of $\mathbf{V}(\hat{\boldsymbol{\beta}})$. Denoting by $\rho_1$ and $\rho_2$ the eigenvalues of $\mathbf{V}(\hat{\boldsymbol{\beta}})$, we deduce from (3.3.1):

$$\rho_1 + \rho_2 = \delta_0 + \delta_1 + (1+q)/pq; \rho_1\rho_2 = \delta_0\delta_1 + (\delta_0/pq) + (\delta_1/p) + (1/pq).$$

Hence the larger of the two eigenvalues is given by

$$\xi(p) = [\delta_0 + \delta_1 + (1+q)/pq]/2 + \sqrt{[\delta_1 - \delta_0 + 1/q]^2 + 4/p^2]}/2$$

When $\delta_0 = \delta_1 = 0$ (the case of fixed effects model), we deduce that $\xi'(p) = 0$ yields : $p_{opt} = 3/5$. For other combinations of (positive) given values of $\delta_0$ and and $\delta_1$ as well, we can minimize $\xi(p)$ in a routine manner. We omit the details. In passing, we note that extensive computations affirm the following : $p_{opt} \geq 3/5$ whenever $\delta_0 \geq \delta_1$ and $1/2 \leq p_{opt} \leq 3/5$ whenever $\delta_0 \leq \delta_1$.

**MV-optimal design**

We start with

$$V(\hat{\beta}_0) = \delta_0 + \frac{1}{p}, \tag{3.3.4}$$

$$V(\hat{\beta}_1) = \delta_1 + \frac{1}{p(1-p)}. \tag{3.3.5}$$

For given $\delta_0$ and $\delta_1$ we have to choose $p$ so as to minimise the larger of the above two expressions. $V(\hat{\beta}_0)$ is monotonically decreasing function of $p$ and $V(\hat{\beta}_1)$ has the minimum at $p = \frac{1}{2}$. After simple calculation, we have

$$V(\hat{\beta}_1) - V(\hat{\beta}_0) > 0 \Leftrightarrow p < 1 - \frac{1}{(\delta_0 - \delta_1)}$$

and $V(\hat{\beta}_0) = V(\hat{\beta}_1)$ exactly when $p = 1 - \frac{1}{(\delta_0 - \delta_1)} = p_0$, say. It follows that we have to distinguish between the two cases:

(i) $(\delta_0 - \delta_1) > 1$: It is easy to see that $V(\hat{\beta}_0) > V(\hat{\beta}_1)$ for $p < p_0$, and $V(\hat{\beta}_1) > V(\hat{\beta}_0)$ for $p > p_0$. Then clearly $\max\{V(\hat{\beta}_0), V(\hat{\beta}_1)\}$ attains its minimum when $p = p_0$ if $p_0 \geq 0.5$; or when $p = 0.5$ if $p_0 \leq 0.5$.

(ii) $(\delta_0 - \delta_1) < 1$: Here $V(\hat{\beta}_0) < V(\hat{\beta}_1)$ for all $p$ and $V(\hat{\beta}_1)$ attains its minimum when $p = 0.5$.

**Remark 3.3.1** It must be noted that the DLG phenomenon continues to hold if instead of $[0,1]$, the design region is $[0,h]$ for some (finite) $h > 0$. In other words, 2-point designs having mass at 0 and $h$ will form a complete class of designs.

With reference to $d_p = \{0, h; p, 1-p\}$, we now have the variance-covariance matrix, revised from (3.3.1) to:

$$\mathbf{V}(\hat{\beta}) = \begin{pmatrix} \delta_0 & 0 \\ 0 & \delta_1 \end{pmatrix} + \frac{1}{Np} \begin{pmatrix} 1 & -1/h \\ -1/h & 1/h^2(1-p) \end{pmatrix}. \quad (3.3.6)$$

For specific optimality, $p_{opt}$ would depend on the criterion used, as also on the choice of $h$. Since no new techniques are involved, we may omit the steps which relate to optimal choice of $p$ for given $h$ for parameter estimation.

### 3.3.2 Quadratic RCR Model

We now turn to the case of quadratic regression with random coefficients attached to each of the fixed coefficients. We borrow the notations from (1.3.11)–(1.3.14) and invoke the DLG phenomenon on the fixed regressions part. Thus it is enough to work with exactly three design points and that too, with the extreme values 0 and 1 and one middle value, say $\lambda$. We thus write:

$$\mathbf{V}_R(\hat{\beta}) = \mathbf{D} + (\mathbf{X}'\mathbf{D}_p\mathbf{X})^{-1} \quad (3.3.7)$$

where

$$\mathbf{X} = \begin{pmatrix} 1 & 0 & 0 \\ 1 & \lambda & \lambda^2 \\ 1 & 1 & 1 \end{pmatrix}, \quad \mathbf{D}_p = \text{Diag}(p, q, r) \quad (3.3.8)$$

and

$$\mathbf{D} = \text{Diag}(\delta_0, \delta_1, \delta_2), \delta_i = V(b_i)/\sigma^2, \ i = 0, 1, 2. \quad (3.3.9)$$

We now proceed to characterize optimal designs *with respect to* specific optimality criteria.

#### Optimal Designs for Full Set of Parameters

#### A-optimal design:

As is already understood, this poses no new problem under RCR models.

#### D-optimal design:

We have to minimize

$$|\mathbf{V}_R(\hat{\beta})| = |\mathbf{X}|^{-2}|\mathbf{D}_p^{-1} + \mathbf{X}'\mathbf{D}\mathbf{X}| = |\mathbf{X}|^{-2}\Psi(p, q, r, \lambda) + \prod \delta_i, \quad (3.3.10)$$

where

$$\begin{aligned}
\Psi(p, q, r, \lambda) &= \frac{a}{p} + \frac{b}{q} + \frac{c}{r} + \frac{d}{pq} + \frac{e}{pr} + \frac{f}{qr} + \frac{1}{pqr}, \\
a &= \delta_0\delta_1(1-\lambda)^2 + \delta_0\delta_2(1-\lambda^2)^2 + \delta_1\delta_2\lambda^2(1-\lambda)^2, \\
b &= \delta_0(\delta_1 + \delta_2), \\
c &= \delta_0\lambda^2(\delta_1 + \delta_2\lambda^2), \\
d &= \delta_0 + \delta_1 + \delta_2, \ e = \delta_0 + \delta_1\lambda^2 + \delta_2\lambda^4, \ f = \delta_0.
\end{aligned}$$

For fixed $\lambda$, the minimization problem in terms of $p, q, r$ results into

$$\begin{aligned} p_{opt} &= a^*/[a^* + b^* + c^*], \\ q_{opt} &= b^*/[a^* + b^* + c^*], \\ r_{opt} &= a^*/[a^* + b^* + c^*], \end{aligned} \tag{3.3.11}$$

where

$$a^* = 1+aqr+dr+eq,\ b^* = 1+bpr+dr+fp,\ c^* = 1+cpq+eq+fp. \tag{3.3.12}$$

Note that the above quantities are implicit functions of $p$, $q$, and $r$. Therefore, for any given value of $\lambda$, we can solve for $p_{opt}$, $q_{opt}$ and $r_{opt}$ by usual iteration method. And then by varying $\lambda$, we can find its optimum value and, hence, finally the optimal design.

It must be noted that while ascertaining optimal value of $\lambda$, we have to take into account the factor $|\mathbf{X}|^{-2}$ in the expression for $|\mathbf{V}_R(\hat{\boldsymbol{\beta}})|$ above. Also all these have to tabulated for various combinations of values of the $\delta$-parameters. Table 3.1 below specifies the nature of $D$-optimal designs for some such combinations of the $\delta$ parameters. Note that $\delta_0 = \delta_1 = \delta_2 = 0$ corresponds to the case of fixed effects quadratic regression model.

**Optimal Designs for Subset of Parameters $\boldsymbol{\beta}^{(2)} = (\beta_1,\ \beta_2)'$**

**A-optimal design**

As is already explained, this poses no new problem under RCR models.

**D-optimal design**

This time we have to minimize $|\mathbf{V}_R(\hat{\boldsymbol{\beta}}^{(2)})|$ which is given by the determinant of $\mathbf{S} + \mathbf{T}$, where $\mathbf{S} = \begin{pmatrix} \delta_1 & 0 \\ 0 & \delta_2 \end{pmatrix}$ and $pqr\lambda^2(1-\lambda)^2\mathbf{T}$ is of the form

$$\begin{pmatrix} q(1-q)\lambda^4 + r(1-r) - 2qr\lambda^2 & q(1-q)\lambda^3 + r(1-r) - qr\lambda(1+\lambda) \\ & q(1-q)\lambda^2 + r(1-r) - 2qr\lambda \end{pmatrix}. \tag{3.3.13}$$

Upon simplification, this yields

$$\begin{aligned} |\mathbf{V}_R(\hat{\boldsymbol{\beta}}^{(2)})| &= \delta_1\delta_2 + [\{1/pqr\} + \delta_1\{(1-\lambda)^2/p + 1/q + \lambda^2/r\} \\ &+ \delta_2\{1-\lambda^2)^2/p + 1/q + \lambda^4/r\}]/\lambda^2(1-\lambda)^2, \end{aligned} \tag{3.3.14}$$

which is of the form $a/p + b/q + c/r + d/pqr$ as a function of $p$, $q$ and $r$.

In the above expressions $a, b, c$ and $d$ are given by

$$\begin{aligned} a &= [\delta_1(1-\lambda)^2 + \delta_2(1-\lambda^2)^2]/\lambda^2(1-\lambda)^2 \\ b &= [\delta_1 + \delta_2]/\lambda^2(1-\lambda)^2 \\ c &= [\delta_1\lambda^2 + \delta_2\lambda^4]/\lambda^2(1-\lambda)^2 \\ d &= 1/\lambda^2(1-\lambda)^2. \end{aligned} \tag{3.3.15}$$

Define

$$a^* = aqr + d,\ b^* = bpr + d \text{ and } c^* = cpq + d. \tag{3.3.16}$$

**Table 3.1.** $D$-optimal designs for full set of parameters in quadratic RCR models: $d_{3:p} = \{0, \lambda_{opt}, 1;\ p_{opt}, q_{opt}, r_{opt}\}$.

| $\delta_0$ | $\delta_1$ | $\delta_2$ | $\lambda_{opt}$ | $p_{opt}$ | $q_{opt}$ | $r_{opt}$ |
|---|---|---|---|---|---|---|
| 0 | 0 | 0 | 0.5 | 0.333 | 0.333 | 0.333 |
| 0 | 0 | 0.5 | 0.495 | 0.350 | 0.346 | 0.304 |
| 0 | 0.5 | 0 | 0.491 | 0.352 | 0.340 | 0.308 |
| 0.5 | 0 | 0 | 0.5 | 0.333 | 0.333 | 0.333 |
| 0 | 0 | 1 | 0.492 | 0.361 | 0.356 | 0.283 |
| 0 | 1 | 0 | 0.485 | 0.366 | 0.345 | 0.289 |
| 1 | 0 | 0 | 0.5 | 0.333 | 0.333 | 0.333 |
| 0 | 0.5 | 0.5 | 0.488 | 0.364 | 0.350 | 0.286 |
| 0.5 | 0 | 0.5 | 0.498 | 0.345 | 0.345 | 0.310 |
| 0.5 | 0.5 | 0 | 0.494 | 0.345 | 0.341 | 0.314 |
| 0.5 | 0.5 | 0.5 | 0.492 | 0.354 | 0.350 | 0.296 |
| 0 | 1 | 1 | 0.481 | 0.383 | 0.358 | 0.259 |
| 1 | 0 | 1 | 0.498 | 0.349 | 0.353 | 0.298 |
| 1 | 1 | 0 | 0.491 | 0.348 | 0.347 | 0.305 |
| 1 | 1 | 1 | 0.490 | 0.360 | 0.360 | 0.280 |
| 5 | 5 | 5 | 0.492 | 0.355 | 0.403 | 0.242 |
| 10 | 10 | 10 | 0.495 | 0.346 | 0.426 | 0.228 |
| 20 | 20 | 20 | 0.497 | 0.338 | 0.447 | 0.215 |
| 30 | 30 | 30 | 0.498 | 0.334 | 0.456 | 0.210 |
| 40 | 40 | 40 | 0.498 | 0.331 | 0.463 | 0.206 |
| 50 | 50 | 50 | 0.499 | 0.330 | 0.467 | 0.203 |
| 1 | 1 | 100 | 0.456 | 0.446 | 0.439 | 0.115 |
| 1 | 100 | 1 | 0.438 | 0.428 | 0.393 | 0.180 |
| 100 | 1 | 1 | 0.504 | 0.328 | 0.361 | 0.311 |

Then, it follows that

$$p_{opt} = a^*/[a^* + b^* + c^*], q_{opt} = b^*/[a^* + b^* + c^*] \text{ and } r_{opt} = c^*/[a^* + b^* + c^*] \tag{3.3.17}$$

Note that the above expressions can be simplified and are seen to be independent of $\lambda^2(1-\lambda)^2$, which occurs in the denominators of the expressions for $a$ to $d$ above.

Anyway, once again, these are implicit functions of $p, q$ and $r$ and we can find an iterative solution for given $\lambda$. Thus minimization of $|\mathbf{V}_R(\boldsymbol{\beta}^{(2)})|$ is, as before, a two-stage process. For given valules of the $\delta$-parameters, we can eventually obtain optimum values of $\lambda$ and hence, of $p, q$ and $r$. Table 3.2 exhibits optimal designs for some combinations of values of the $\delta$-parameters.

**MV-optimal design**

The pupose is to characterize an optimal design in the sense of providing

**Table 3.2.** $D$-optimal designs for subset of parameters in quadratic RCR models: $d_{3:p} = \{0, \lambda_{opt}, 1; p_{opt}, q_{opt}, r_{opt}\}$.

| $\delta_1$ | $\delta_2$ | $\lambda_{opt}$ | $p_{opt}$ | $q_{opt}$ | $r_{opt}$ |
|---|---|---|---|---|---|
| 0 | 0 | 0.5 | 0.333 | 0.333 | 0.333 |
| 0 | 0.5 | 0.503 | 0.334 | 0.341 | 0.325 |
| 0.5 | 0 | 0.5 | 0.329 | 0.342 | 0.329 |
| 0 | 1 | 0.506 | 0.334 | 0.349 | 0.317 |
| 1 | 0 | 0.5 | 0.325 | 0.350 | 0.325 |
| 1 | 1 | 0.505 | 0.327 | 0.361 | 0.312 |
| 5 | 5 | 0.509 | 0.319 | 0.412 | 0.269 |
| 10 | 10 | 0.509 | 0.316 | 0.437 | 0.247 |
| 20 | 20 | 0.508 | 0.314 | 0.457 | 0.228 |
| 30 | 30 | 0.507 | 0.314 | 0.466 | 0.220 |
| 40 | 40 | 0.506 | 0.313 | 0.472 | 0.215 |
| 50 | 50 | 0.505 | 0.313 | 0.475 | 0.211 |
| 100 | 100 | 0.504 | 0.313 | 0.483 | 0.204 |

the least value of $\max\{V_R(\hat{\beta}_1), V_R(\hat{\beta}_2)\}$. As is well known, this is attained when $V_R(\hat{\beta}_1) = V_R(\hat{\beta}_2)$. From the expression for $\mathbf{V}_R(\hat{\boldsymbol{\beta}}^{(2)})$, it follows that this holds if $V_F(\hat{\beta}_1) + \delta_1 = V_F(\hat{\beta}_2) + \delta_2$, which simplifies to

$$2/p - [\lambda(1+\lambda)](1/r + 1/p) = (\delta_2 - \delta_1)\lambda(1-\lambda) \tag{3.3.18}$$

whence

$$\begin{aligned} & 1/p = a + b/r, a = [(\delta_2 - \delta_1)\lambda(1-\lambda)]/[2 - \lambda(1+\lambda)], \\ \text{and} \quad & b = \lambda(1+\lambda)/[2 - \lambda(1+\lambda)]. \end{aligned} \tag{3.3.19}$$

Subject to this condition, we have to minimize the common value of $V_R(\hat{\beta}_1)$ and $V_R(\hat{\beta}_2)$. We start with $V_R(\hat{\beta}_1)$ and proceed to minimize it, leaving aside the term $\delta_1$.

Upon simplification, the term to be minimized turns out to be

$$\begin{aligned} \Psi(\lambda, p, q, r) &= [(1+\lambda)^2/p\lambda^2] + [1/q\lambda^2(1-\lambda)^2] + [\lambda^2/r(1-\lambda)^2] \\ &= a^*/p + b^*/q + c^*/r, \end{aligned} \tag{3.3.20}$$

say. Using Lagranges' multipliers corresponding to the restrictions

$$p + q + r = 1 \text{ and } p = a + b/r, \tag{3.3.21}$$

we end up with the following equations for $p$, $q$ and $r$, for given $\lambda$:

$$p_{opt} = \sqrt{\{a^* + \Lambda\}}/[\sqrt{\{a^* + \Lambda\}} + \sqrt{b^*} + \sqrt{\{c^* - \Lambda b\}}]$$

$$q_{opt} = \sqrt{b}/[\sqrt{\{a^* + \Lambda\}} + \sqrt{b^*} + \sqrt{\{c^* - \Lambda b\}}]$$

$$r_{opt} = \sqrt{\{c^* + \Lambda b\}}/[\sqrt{\{a^* + \Lambda\}} + \sqrt{b^*} + \sqrt{\{c^* - \Lambda b\}}],\qquad(3.3.22)$$

where

$$\Lambda = [c^* - a^*\{a + b/r\}^2]/[b^*\{a + b/r\}^2]. \qquad (3.3.23)$$

Thus, in effect, we have to solve recursively for $p_{opt}, q_{opt}$ and $r_{opt}$ for any given $\lambda$. However, the optimum value of $\lambda$ is to be ascertained by minimizing

$$\Psi(\lambda) = [a^*/\sqrt{\{a^* + \Lambda\}} + \sqrt{b^*} + c^*/\sqrt{\{c^* - \Lambda b\}}][\sqrt{\{a^* + \Lambda\}} + \sqrt{b^*} + \sqrt{\{c^* - \Lambda b\}}]. \qquad (3.3.24)$$

Numerical computations will now enable us to study the nature of optimal designs for various combinations of the values of the $\delta$-parameters. Table 3.3 below exhibits $MV$-optimal designs for some combinations of the $\delta$-parameters.

**Table 3.3.** $MV$-optimal designs for subset of parameters in quadratic RCR models: $d_{3:p} = \{0, \lambda_{opt}, 1; p_{opt}, q_{opt}, r_{opt}\}$.

| $\delta_1$ | $\delta_2$ | $\lambda_{opt}$ | $p_{opt}$ | $q_{opt}$ | $r_{opt}$ |
|---|---|---|---|---|---|
| 0 | 0 | 0.489 | 0.348 | 0.493 | 0.159 |
| 0 | 0.5 | 0.488 | 0.345 | 0.493 | 0.162 |
| 0.5 | 0 | 0.490 | 0.351 | 0.494 | 0.155 |
| 0.5 | 0.5 | 0.489 | 0.348 | 0.493 | 0.159 |
| 0 | 1 | 0.487 | 0.342 | 0.493 | 0.165 |
| 1 | 0 | 0.492 | 0.354 | 0.494 | 0.152 |
| 1 | 1 | 0.489 | 0.348 | 0.493 | 0.159 |
| 5 | 5 | 0.489 | 0.348 | 0.493 | 0.159 |
| 5 | 25 | 0.441 | 0.243 | 0.510 | 0.247 |
| 25 | 5 | 0.514 | 0.322 | 0.492 | 0.186 |
| 5 | 50 | 0.396 | 0.185 | 0.543 | 0.272 |
| 50 | 5 | 0.540 | 0.201 | 0.507 | 0.292 |
| 5 | 100 | 0.341 | 0.138 | 0.583 | 0.279 |
| 100 | 5 | 0.489 | 0.117 | 0.551 | 0.332 |

# 3.4 Optimal Designs for Prediction in RCR Models

## 3.4.1 Introduction

We now refer to model (1.2.1) of Chapter 1. Both the linear regression and the quadratic regression are special cases of it. We have shown that for this model

$$\mathbf{V}_R(\hat{\boldsymbol{\beta}}) = \mathbf{D} + \mathbf{V}_F(\hat{\boldsymbol{\beta}}) = \mathbf{D} + \mathcal{I}_F^{-1},$$

where $\mathcal{I}_F = \mathbf{X}'\mathbf{D}_p\mathbf{X}$ is the information matrix of the design (vide (1.3.15)). Looking at the above expression, it is clear that the inclusion of the term $\mathbf{D}$ in $\mathbf{V}_R(\hat{\boldsymbol{\beta}})$ does *not* alter the minimization problem in terms of $\mathbf{X}$ and $d_p$. Hence the *same* design minimizes the prediction criteria considered below for the fixed regression coefficients case or for the random regression coefficients case.

## 3.4.2 First-Degree RCR Model

We now address the more challenging problem of prediction of future observations beyond the design region, i.e., for values of $x$ in

$$\mathcal{T}^+ = \{x : h < x \leq H\}. \tag{3.4.1}$$

The point predictor of $Y_x$ is given by

$$\hat{Y}_x = \hat{\beta}_0 + \hat{\beta}_1 x. \tag{3.4.2}$$

Instead of predicting $Y_x$ at pre-specified points in $\mathcal{T}^+$, we wish to predict $Y_x$ at an arbitrary point in it. Hence we consider the variance function

$$\begin{aligned} V(\hat{Y}_x) &= (1,\ x)\mathbf{V}(\hat{\boldsymbol{\beta}})(1,\ x)' \\ &= V(\hat{\beta}_0) + x^2 V(\hat{\beta}_1) + 2x\mathrm{Cov}(\hat{\beta}_0, \hat{\beta}_1), \end{aligned} \tag{3.4.3}$$

whose elements are to come from (3.1.4). Since $x$ varies in (3.4.1), we consider two options in choosing the measure of goodness of prediction.

**The Integrated Mean Square Error (IMSE) Criterion**
In this section we define IMSE-optimum design as one which minimizes

$$\bar{V} = IMSE = \int_{\mathcal{T}^+} V(\hat{Y}_x) w(x)\, dx, \tag{3.4.4}$$

where $w(x)$ is a suitably chosen weight function in $(h, H]$. For uniform weight function, the weight can be ignored in (3.4.4). Below, we work with uniform weight function.

Since in prediction as well as in estimation we are dealing with the dispersion matrix, it follows that the DLG phenomenon still applies and hence it suffices to restrict attention to the class of 2-point designs concentrated at the extremes. All we need to determine now is the value of $p$ (the weight at point 0) with reference to $d_{2:p} = (0, h\ ;\ p,\ 1-p)$.

Routine calculations yield

$$\bar{V} = \frac{V(\hat{\beta}_0) + V(\hat{\beta}_1)\{H^2 + Hh + h^2\}}{3} + (H+h)\mathrm{Cov}(\hat{\beta}_0, \hat{\beta}_1). \tag{3.4.5}$$

From (3.3.1), it turns out that minimisation of (3.4.5) can be done irrespective of the values of $\delta_0$ and $\delta_1$. Writing

$$\begin{aligned} \phi &= h/H, \\ A &= (1 + \phi + \phi^2)/3\phi, \end{aligned} \tag{3.4.6}$$

the optimum value of $p$ is given by

$$p_{\bar{V}} = \frac{\sqrt{A-1}}{\sqrt{A-1}+\sqrt{A}} \tag{3.4.7}$$

$$\approx \frac{1-\phi}{(1-\phi)+\sqrt{1+\phi+\phi^2}}. \tag{3.4.8}$$

**The Minimax Criterion**

Here we choose the design which minimizes

$$V^* = \max_{x \in T^+} V(\widehat{Y}_x). \tag{3.4.9}$$

Since $V(\widehat{Y}_x)$ is convex in $x$, its maximum is reached at one of the end points $h$ or $H$. Further, it can be shown (Liski *et al.* 1998) that

$$V(\widehat{Y}_x)|_{x=h} < V(\widehat{Y}_x)|_{x=H}. \tag{3.4.10}$$

Thus we have to minimise the variance of the predictor at the point $x = H$. The optimum value of $p$ is given by the same form as in (3.4.7) with $A \approx 1/[\phi(2-\phi)]$. This yields

$$P_{V^*} \approx \frac{1-\phi}{2-\phi}. \tag{3.4.11}$$

In the following tables we show approximate values of $p_{opt}$ for both criteria.

**Table 3.4.** Optimal (approx.) values of $p$ according to IMSE criterion.

| $\phi$ | 0.1 | 0.2 | 0.3 | 0.4 | 0.5 | 0.6 | 0.7 | 0.8 | 0.9 |
|---|---|---|---|---|---|---|---|---|---|
| $p_{\bar{V}}$ | 0.46 | 0.42 | 0.37 | 0.32 | 0.27 | 0.22 | 0.17 | 0.11 | 0.06 |

**Table 3.5.** Optimal (approx.) values of $p$ according to minimax criterion.

| $\phi$ | 0.1 | 0.2 | 0.3 | 0.4 | 0.5 | 0.6 | 0.7 | 0.8 | 0.9 |
|---|---|---|---|---|---|---|---|---|---|
| $p_{V^*}$ | 0.47 | 0.44 | 0.41 | 0.38 | 0.33 | 0.29 | 0.23 | 0.17 | 0.09 |

From the above Tables it is clear that with respect to both criteria the optimum value of $p$ decreases with increasing values of $\phi$. This is quite understandable, since in such cases there will be increasing importance of the extreme point $h$. For small values of $\phi$, the extreme point $h$ loses its relative importance, though it has always a weight greater than zero.

We re-iterate that in the above, for the IMSE criterion as well as for the minimax criterion, the choice of the optimal designs does *not* depend on the fixed or random nature of the regression coefficients.

### 3.4.3 Quadratic RCR Model

We now consider the problem of prediction for the quadratic RCR model which is a natural extension of (3.2.1). As before, the point predictor of $Y_x$ is given by

$$\hat{Y}_x = \hat{\beta}_0 + \hat{\beta}_1 x + \hat{\beta}_2 x^2, \tag{3.4.12}$$

where $\hat{\beta}_0$, $\hat{\beta}_1$ and $\hat{\beta}_2$ are the least squares estimators of $\beta_0$, $\beta_1$ and $\beta_2$, respectively.

We want to predict $Y_x$ at an arbitrary point in $\mathcal{T}^+$. Consider the variance function

$$V(\hat{Y}_x) = (1,\ x,\ x^2)V(\hat{\boldsymbol{\beta}})(1,\ x,\ x^2)', \tag{3.4.13}$$

where $\hat{\boldsymbol{\beta}} = (\hat{\beta}_0,\ \hat{\beta}_1,\ \hat{\beta}_2)'$. Since $x$ varies in $\mathcal{T}^+$, we use the IMSE as the criterion in selecting a design. One can of course, pursue the minimax approach. However, since $V(\hat{Y}_x)$ given by (3.4.13) is a polynomial in fourth degree and the coefficients involve the design parameters, it might be difficult to find a minimax design in a closed form.

Now to find a design using IMSE criterion, we have to minimize, with the uniform weight function,

$$\begin{aligned} \text{IMSE} &= \int_h^H V(\hat{Y}_x)dx \\ &= \text{tr}[\mathbf{V}(\hat{\boldsymbol{\beta}})\mathbf{B}] \end{aligned} \tag{3.4.14}$$

where $\mathbf{B} = \int_h^H (1,\ x,\ x^2)'(1,\ x,\ x^2)dx$. Now for an RCR model

$$\mathbf{V}(\hat{\boldsymbol{\beta}}) = \mathbf{D} + (\mathbf{X}'\mathbf{D}_p\mathbf{X})^{-1} \tag{3.4.15}$$

and hence IMSE given by (3.4.14) can be expressed as

$$\text{IMSE} = \text{tr}\ \{[\mathbf{D} + (\mathbf{X}'\mathbf{D}_p\mathbf{X})^{-1}]\mathbf{B}\}. \tag{3.4.16}$$

Since the first part in (3.4.16) is independent of the design, the minimization of (3.4.16) with respect to the design is equivalent to the minimization of

$$\text{tr}[(\mathbf{X}'\mathbf{D}_p\mathbf{X})^{-1}\mathbf{B}] = \text{tr}(\mathcal{I}^{-1}\mathbf{B}). \tag{3.4.17}$$

Thus the optimum design under IMSE criterion for the fixed effects and the RCR model is same. This is true for any polynomial regression model. Now, for a quadratic regression model, $\mathbf{B}$ reduces to

$$\mathbf{B} = \begin{pmatrix} H(1-\phi) & \frac{H^2}{2}(1-\phi^2) & \frac{H^3}{3}(1-\phi^3) \\ & \frac{H^3}{3}(1-\phi^3) & \frac{H^4}{4}(1-\phi^4) \\ & & \frac{H^5}{5}(1-\phi^5) \end{pmatrix}, \tag{3.4.18}$$

where $\phi = h/H$. Since the reduced criterion function (3.4.17) is a function of the information matrix, we can use the DLG phenomenon and restrict to the class of 3-point designs: $d_3[0, \lambda, h; p, q, r]$. For such a $d_3$, the information matrix $\mathcal{I}$ is given by

$$\mathcal{I} = \begin{pmatrix} 1 & q\lambda + rh & q\lambda^2 + rh^2 \\ & q\lambda^2 + rh^2 & q\lambda^3 + rh^3 \\ & & q\lambda^4 + rh^4 \end{pmatrix}.$$

After a little algebra, it can be shown that $\text{tr}(\mathcal{I}^{-1}\mathbf{B})$ can be expressed as

$$\text{tr}(\mathcal{I}^{-1}\mathbf{B}) = \frac{A}{p} + \frac{B}{q} + \frac{C}{r}, \tag{3.4.19}$$

where

$$\begin{aligned} A &= H[(1-\phi) - \frac{1-\phi^2}{\phi\xi}\frac{[1-\xi(1+\xi)+\xi^3]}{(1-\xi)^2} + \frac{2}{3}\frac{1-\phi^3}{\phi^2\xi} + \frac{1-\phi^3}{3\phi^2\xi^2}(1+\xi)^2 \\ &+ \frac{(1-\phi^5)}{5\phi^4\xi^2} - \frac{(1-\phi^4)[1-\xi(1+\xi)+\xi^3]}{2\phi^3\xi^2(1-\xi)^2} \\ B &= \frac{H}{\phi^2\xi^2(1-\xi)^2}\left[\frac{(1-\phi^3)}{1} - \frac{(1-\phi^4)}{2\phi} + \frac{(1-\phi^5)}{5\phi^2}\right] \\ C &= \frac{H}{\phi^2(1-\xi)^2}\left[\frac{\xi^2(1-\phi^3)}{3} - \frac{\xi(1-\phi^4)}{2\phi} + \frac{(1-\phi^5)}{\phi^2}\right] \end{aligned} \tag{3.4.20}$$

with $\phi = h/H, \xi = \lambda/h; \phi, \xi \in (0,1)$.

From (3.4.19) we get by Cauchy - Schwartz inequality that

$$\text{tr}\, \mathcal{I}^{-1}\mathbf{B} \geq (\sqrt{A} + \sqrt{B} + \sqrt{C})^2 \tag{3.4.21}$$

and "=" is attained at

$$p = p^* = \frac{\sqrt{A}}{\sqrt{A}+\sqrt{B}+\sqrt{C}}, \quad q = q^* = \frac{\sqrt{B}}{\sqrt{A}+\sqrt{B}+\sqrt{C}}$$

and

$$r = r^* = \frac{\sqrt{C}}{\sqrt{A}+\sqrt{B}+\sqrt{C}}. \tag{3.4.22}$$

Finally, we have to minimize the right hand side of (3.4.21) with respect to $\xi$ in $(0,1)$ for given values of $\phi$ in $(0,1)$. Once the optimum value of $\xi = \xi^*$ is obtained, one can find the values of $p^*$, $q^*$ and $r^*$ from (3.4.22). In Table 3.6 we display optimal designs for some selected values of $\phi = h/H$. Note that $\lambda_{opt}$ is dependent on $\xi_{opt}$ and $h$.

## 3.5 Optimal Designs for Inverse Prediction in RCR Models

We now address the problem of 'Inverse Prediction', which requires specification of an optimal design for determining the value of the regressor $x$

**Table 3.6.** Optimal regression designs over $[0, h]$ for prediction over $[h, H]$ in a quadratic model $d = (0,\ \xi_{opt},\ h;\ p_{opt},\ q_{opt},\ r_{opt})$.

| $h/H$ | $\xi_{opt}$ | $p_{opt}$ | $q_{opt}$ | $r_{opt}$ |
|---|---|---|---|---|
| 0.05 | 0.4358 | 0.212 | 0.383 | 0.405 |
| 0.10 | 0.4313 | 0.203 | 0.385 | 0.412 |
| 0.15 | 0.4268 | 0.193 | 0.385 | 0.422 |
| 0.20 | 0.4225 | 0.181 | 0.388 | 0.431 |
| 0.30 | 0.4146 | 0.154 | 0.407 | 0.439 |
| 0.40 | 0.4085 | 0.124 | 0.438 | 0.438 |
| 0.50 | 0.4041 | 0.093 | 0.477 | 0.430 |
| 0.60 | 0.4011 | 0.066 | 0.517 | 0.417 |
| 0.70 | 0.3989 | 0.043 | 0.553 | 0.404 |
| 0.80 | 0.3972 | 0.025 | 0.585 | 0.390 |
| 0.90 | 0.3958 | 0.011 | 0.611 | 0.378 |

when a given value of the (mean) response is to be achieved through the regression equation. We will first deal with the case of linear and quadratic regression with fixed coefficients. Then we will argue that the case of random coefficients does not pose any new problem. In fact, the results are the same irrespective of the nature of of the regression coefficients as being fixed or random.

### 3.5.1 First-Degree Regression Model

We start with the first-degree regression equation

$$E(Y_x) = \beta_0 + \beta_1 x. \tag{3.5.1}$$

Let $\eta_0$ denote the mean response to be achieved, so that the problem is one of determining $x_0$ for which

$$x_0 = \frac{\eta_0 - \beta_0}{\beta_1}. \tag{3.5.2}$$

Since $\beta_0$ and $\beta_1$ are assumed to be unknown, we can replace them by their least squares estimates to obtain

$$\hat{x}_0 = \frac{\eta_0 - \hat{\beta}_0}{\hat{\beta}_1}. \tag{3.5.3}$$

We now note that $x_0$ is a non-linear function of $\beta_0$ and $\beta_1$. Hence, any measure of accuracy of $\hat{x}_0$ will depend on the performance of the estimates of these unknown parameters.

We assume that from past experiences, we have some prior knowledge concerning these unknown parameters. This can be expressed in the form of a *prior distribution* for $\beta_0$ and $\beta_1$. Next, we assume that $x_0 = (\eta_0 - \beta_0)/\beta_1$ is distributed independently of $1/\beta_1^2$. Moreover, we assume that

$$\mathcal{E}(x_0) = \mu, \quad \mathcal{V}(x_0) = v. \tag{3.5.4}$$

Here $\mathcal{E}$ and $\mathcal{V}$ are the expectation and variance with reference to the prior distribution of $x_0$. From a practical point of view, it appears that the response values can be made available at the beginning of the range of the values of the regressor $x$. As earlier, we therefore take the experimental region as

$$0 \leq x \leq h. \tag{3.5.5}$$

We now make the following transformation:

$$x^* = \frac{x - \mu}{\sqrt{v}}, \qquad x_0^* = \frac{x_0 - \mu}{\sqrt{v}}. \tag{3.5.6}$$

Next we rewrite the model (3.5.1) as

$$E(Y_x|x^*) = \beta_0^* + \beta_1^* x^*, \tag{3.5.7}$$

where $\beta_0^* = \beta_0 + \beta_1\mu$ and $\beta_1^* = \beta_1\sqrt{v}$. Moreover,

$$\mathcal{E}(x_0^*) = 0, \qquad \mathcal{V}(x_0^*) = 1 \tag{3.5.8}$$

and the experimental region (3.5.5) is transformed to

$$L \leq x^* \leq U \tag{3.5.9}$$

with

$$L = \frac{-\mu}{\sqrt{v}}, \qquad U = \frac{h - \mu}{\sqrt{v}}. \tag{3.5.10}$$

In what follows, we work with the set-up (3.5.7)–(3.5.8) and drop the asterisks in $x^*$, $x_0^*$, $\beta_0^*$ and $\beta_1^*$ throughout. Since $x_0$ given by (3.5.2) is non-linear in the parameters, we apply the $\delta$ method and a large sample approximation to find $V(\hat{x}_0)$:

$$V(\hat{x}_0) \approx \left(\frac{\partial x_0}{\partial \beta_0}, \frac{\partial x_0}{\partial \beta_1}\right) \mathbf{V}(\hat{\beta}) \left(\frac{\partial x_0}{\partial \beta_0}, \frac{\partial x_0}{\partial \beta_1}\right)', \tag{3.5.11}$$

where $\mathbf{V}(\hat{\beta})$ is the dispersion matrix of $\hat{\beta} = (\hat{\beta}_0, \hat{\beta}_1)'$, the least squares estimator of $\beta_0$ and $\beta_1$ based on the continuous design

$$d_n = (x_1, x_2, \ldots, x_n; \; p_1, p_2, \ldots, p_n).$$

The problem is that of finding an optimal $d_n$ in order to estimate $x_0$ most efficiently. As earlier, we work with the continuous set-up. It is easy to see that

$$\left(\frac{\partial x_0}{\partial \beta_0}, \frac{\partial x_0}{\partial \beta_1}\right) = -(1, x_0)/\beta_1,$$

so that (3.5.11) reduces to

$$V(\hat{x}_0) = \frac{(1,\ x_0)\mathbf{V}(\hat{\beta})(1,\ x_0)'}{\beta_1^2}. \tag{3.5.12}$$

Taking expectation of (3.5.12) with respect to the prior distribution, we have

$$\begin{aligned} \mathcal{E}[V(\hat{x}_0)] &= \mathcal{E}\left(\frac{1}{\beta_1^2}\right) \times \mathrm{tr}\{\mathbf{V}(\hat{\beta}) \times \mathcal{E}([(1,\ x_0)'(1,\ x_0)])\} & (3.5.13) \\ &= \mathcal{E}\frac{1}{\beta_1^2(\mu_2 - \mu_1^2)}(1 + \mu_2). & (3.5.14) \end{aligned}$$

Here $\mu_1$ and $\mu_2$ are the moments of the design points defined earlier. Since (3.5.13) is a function of the information matrix, we can use the DLG phenomenon to restrict ourselves to an extreme-points design i.e., a design which assigns all its mass at the points $L$ and $U$. It can be shown that the design which minimizes (3.5.13) assigns mass $p_{opt}$ at 0 and mass $q_{opt}$ at $h$ (in the original experimental domain$[0, h]$) where

$$p_{opt} = \frac{-(1+U^2) + \sqrt{(1+U^2)(1+L^2)}}{L^2 - U^2}. \tag{3.5.15}$$

The value of (3.5.13) at (3.5.15) is given by

$$\mathcal{E}[(V(\hat{x}_0)] = \mathcal{E}\left(\frac{1}{\beta_1^2}\right)\frac{1+\mu_2}{\mu_2}. \tag{3.5.16}$$

As an illustration, consider the experimental region $[0.0, 5.0]$ with $\mu = 20$ and $v = 4$. Then in terms of the transformed variable, we get $L = -10.0$ and $U = -7.5$. The minimisation problem yields $p_{opt} = 0.4295$ and the optimal design is given by:

| | | |
|---|---|---|
| Point : | 0.0 | 5.0, |
| Mass : | 0.43 | 0.57. |

### 3.5.2 Quadratic Regression

We now consider quadratic regression (in the transformed experimental domain)

$$E(Y_x) = \eta_x = \beta_0 + \beta_1 x + \beta_2 x^2,\ \beta_2 < 0. \tag{3.5.17}$$

The problem is then to estimate $[x_0 = \frac{-\beta_1 \pm \sqrt{\beta_1^2 - 4\beta_2(\beta_0 - \eta_0)}}{2\beta_2}]$ for a given $\eta_0$ when all the parameters in the model (3.5.17) are unknown. As before, we use the plug-in estimate

$$\hat{x}_0 = \frac{-\hat{\beta}_1 \pm \sqrt{\hat{\beta}_1^{\,2} - 4\hat{\beta}_2(\hat{\beta}_0 - \eta_0)}}{2\hat{\beta}_2}, \tag{3.5.18}$$

where $\hat{\beta}_0$, $\hat{\beta}_1$ and $\hat{\beta}_2$ are the ordinary least squares estimates of the regression coefficients. Using the $\delta$ method once again it can be shown that

$$V(\hat{x}_0) \approx \frac{\mathbf{f}'(x_0)\mathbf{V}(\hat{\boldsymbol{\beta}})\mathbf{f}(x_0)}{(\beta_1 + 2\beta_2 x_0)^2}, \tag{3.5.19}$$

where $\mathbf{f}(x_0) = (1,\ x_0,\ x_0^2)'$ and $\mathbf{f}'(x_0)$ refers to the transpose of $\mathbf{f}(x_0)$. Further, $\hat{\boldsymbol{\beta}}$ denotes the LSE of $\boldsymbol{\beta} = (\beta_0,\ \beta_1,\ \beta_2)'$. Substituting $\mathbf{V}(\hat{\boldsymbol{\beta}}) = \mathbf{M}^{-1}$ into (3.5.19) yields

$$V(\hat{x}_0) \approx \frac{\mathbf{f}'(x_0)\mathbf{M}^{-1}\mathbf{f}(x_0)}{(\beta_1 + 2\beta_2 x_0)^2}. \tag{3.5.20}$$

To find an optimum design we have to minimise (3.5.20). As in the linear case, this also involves unknown parameters $\beta_0, \beta_1, \beta_2$. Again we assume that we have some prior information on these parameters which can be stated in the form of a prior distribution. Computations of the optimal design are greatly simplified if we assume that for this prior distribution

$$\mathcal{E}\left(\frac{x_0^i}{(\beta_1 + 2\beta_2 x_0)^2}\right) = \mathcal{E}(x_0^i) \times \mathcal{E}\left(1/(\beta_1 + 2\beta_2 x_0)^2\right),\ i = 1, 2, 3, 4.$$

Moreover, besides (3.5.8), we further assume that

$$\mathcal{E}(x_0^{*3}) = 0 \text{ and } \mathcal{E}(x_0^{*4}) = \tau,$$

where $x_0^* = \frac{x_0 - \mu}{\sqrt{v}}$. Taking expectation with respect to the prior distribution of $x_0$, we obtain

$$\begin{aligned} \mathcal{E}([V(\hat{x}_0)]) &= \mathcal{E}\left\{\mathbf{f}'(x_0)\mathbf{M}^{-1}\mathbf{f}(x_0)/(\beta_1 + 2\beta_2 x_0)^2\right\} \\ &= \mathcal{E}\left\{\mathbf{f}'(x_0)\mathbf{M}^{-1}\mathbf{f}(x_0)\right\} \times \mathcal{E}\left\{\frac{1}{(\beta_1 + 2\beta_2 x_0)^2}\right\}. \end{aligned} \tag{3.5.21}$$

Thus minimising $\mathcal{E}([V(\hat{x}_0)])$ with respect to the design is equivalent to minimising $G(\mathbf{M}^{-1})$, where

$$G(\mathbf{M}^{-1}) = \mathrm{tr}\left\{\mathbf{M}^{-1}\mathcal{E}\left(\mathbf{f}(x_0)\mathbf{f}'(x_0)\right)\right\} = \mathrm{tr}\mathbf{M}^{-1}\mathbf{B}_0, \tag{3.5.22}$$

$\mathbf{B}_0 = \mathcal{E}((\mathbf{f}(x_0)\mathbf{f}'(x_0)))$ being a positive definite matrix.

From our observations in Chapter 1, it is clear that for this minimization problem, it is enough to restrict our attention to the class of 3-point designs of the type: $(L,\ s,\ U;\ p,\ q,\ r)$. For any design of this type, we can write $\mathbf{M}$ as

$$\mathbf{M} = p\mathbf{f}(L)\mathbf{f}'(L) + q\mathbf{f}(s)\mathbf{f}'(s) + r\mathbf{f}(U)\mathbf{f}'(U). \tag{3.5.23}$$

Let $\gamma(L)$ be a vector orthogonal to $\mathbf{f}(s)$ and $\mathbf{f}(U)$. Then it is readily seen that

$$\mathbf{M}^{-1}\mathbf{f}(L) = \gamma(L)/\left\{p\mathbf{f}'(L)\gamma(L)\right\}.$$

Similarily, we define $\gamma(s)$ and $\gamma(U)$. It follows that

$$\begin{aligned} \mathbf{M}^{-1}\mathbf{f}(s) &= \gamma(s)/\left\{q\mathbf{f}'(s)\gamma(s)\right\}, \\ \mathbf{M}^{-1}\mathbf{f}(U) &= \gamma(U)/\left\{r\mathbf{f}'(U)\gamma(U)\right\}. \end{aligned}$$

We now present a Lemma which will help us in our search for an optimal design.

**Lemma 3.5.1**

$$G(\mathbf{M}^{-1}) = tr(\mathbf{M}^{-1}\mathbf{B}_0) = \frac{A_L}{p} + \frac{A_s}{q} + \frac{A_U}{r}, \qquad (3.5.24)$$

*where*

$$\begin{aligned} A_L &= \gamma'(L)\mathbf{B}_0\gamma(L)/[\mathbf{f}'(L)\gamma(L)]^2, \\ A_s &= \gamma'(s)\mathbf{B}_0\gamma(s)/[\mathbf{f}'(s)\gamma(s)]^2, \\ A_U &= \gamma'(U)\mathbf{B}_0\gamma(U)/[\mathbf{f}'(U)\gamma(U)]^2. \end{aligned} \qquad (3.5.25)$$

**Proof.** See Appendix A 3.5.1.

The optimal design is to be obtained by minimizing trace $(\mathbf{M}^{-1}\mathbf{B}_0)$. From its form given by Lemma 3.5.1, it is easy to see that the minimum subject to restrictions $0 < p, q, r < p + q + r = 1$ is obtained when

$$\begin{aligned} p &= p_{opt} &= \sqrt{A_L}/\left(\sqrt{A_L} + \sqrt{A_s} + \sqrt{A_U}\right), \\ q &= q_{opt} &= \sqrt{A_s}/\left(\sqrt{A_L} + \sqrt{A_s} + \sqrt{A_U}\right), \\ r &= r_{opt} &= \sqrt{A_U}/\left(\sqrt{A_L} + \sqrt{A_s} + \sqrt{A_U}\right). \end{aligned}$$

The minimum value of $G(\mathbf{M}^{-1})$ is then given by

$$G(\mathbf{M}^{-1}) = \left(\sqrt{A_L} + \sqrt{A_S} + \sqrt{A_U}\right)^2. \qquad (3.5.26)$$

Finally, we have to determine the only unknown quantity $s$ which lies in $(L, U)$. An exact analytic solution for $s$ is quite difficult to attain. Liski *et al.* (1997) reported extensive numerical computations which indicate that the value of $s$ do not differ much from $(L + U)/2$. We therefore take the mid-point viz., $(L + U)/2$ and derive the nature of the optimal 3-point design. For given $L$, $U$ and $\tau$ (the fourth order moment of $x_0$) we can now find $p_{opt}$, $q_{opt}$ and $r_{opt}$ from the expressions given above.

## Numerical computations and major findings

Extensive numerical computations show that the optimal 3-point design is quite robust against possible variation in the values of $\tau$ (cf. Liski *et al.* 1997). However, this depends on the extreme values of the reduced experimental domain. In the following table we set out some findings from Liski *et al.* (1997).

**Table 3.7.** Optimal weight $p$, $q$, $r$ for some combinations of $L$, $U$ and $\tau$.

| | | $\tau = 2$ | | | $\tau = 3$ | | |
|---|---|---|---|---|---|---|---|
| L | U | $p(0)$ | $r(h/2)$ | $q(h)$ | $p(0)$ | $r(h/2)$ | $q(h)$ |
| −2 | −1 | 0.1804 | 0.5227 | 0.2969 | 0.1841 | 0.5215 | 0.2944 |
| −3 | −1 | 0.0600 | 0.5233 | 0.4167 | 0.0615 | 0.5234 | 0.4151 |
| −3 | −2 | 0.0843 | 0.7303 | 0.1854 | 0.0851 | 0.7282 | 0.1867 |
| −4 | −3.5 | 0.0685 | 0.8417 | 0.0898 | 0.0691 | 0.8403 | 0.0906 |
| −5 | −4.8 | 0.0803 | 0.8323 | 0.0874 | 0.0816 | 0.8307 | 0.0877 |

## Appendix A 3.2.1

First we rewrite $\tilde{p_2} + \tilde{p_3}$ using the expressions from (3.2.6). Then we have

$$\tilde{p_2} + \tilde{p_3} = \sum_{i=1}^{3} x_i p_i + \left[\sum_{i=1}^{3} x_i(1-x_i)p_i\right]^2 \left[\sum_{i=1}^{3} x_i^2(1-x_i)p_i\right].$$

Now the inequality $\tilde{p_2} + \tilde{p_3} < 1$ is equivalent to

$$\frac{[\sum x_i(1-x_i)p_i]^2}{[\sum x_i^2(1-x_i)p_i]} < 1 - \sum p_i x_i = \sum p_i(1-x_i),$$

which is the same as

$$\left[\sum x_i(1-x_i)p_i\right]^2 < \left[\sum x_i^2(1-x_i)p_i\right]\left[\sum (1-x_i)p_i\right].$$

This last inequality now follows from the Cauchy–Schwartz inequality.

## Appendix A 3.2.2

We first re-write the expression for $\tilde{\mu_{33}} - \mu_{33}$, using (3.2.4) for $\mu_{12}$. We have to prove

$$\tilde{\mu_{33}} - \mu_{33} = \tilde{p_2}\tilde{x}^4 + \tilde{p_3} - \mu_{33} > 0. \tag{3.5.27}$$

Next using (3.2.4), we may write the inequality

$$\tilde{p_2}\tilde{x}^3(1-\tilde{x}) < \sum_{i=1}^{3} x_i^3(1-x_i)p_i, \tag{3.5.28}$$

which is equivalent to (3.5.27). Substituting the solutions for $\tilde{x}$ and $\tilde{p_2}$ from (3.2.5) and (3.2.6) respectively, we may write (3.5.28) as

$$\left[\sum_{i=1}^{3} x_i^3(1-x_i)p_i\right]\left[\sum_{i=1}^{3} x_i(1-x_i)p_i\right] > \left[\sum_{i=1}^{3} x_i(1-x_i)p_i\right]^2,$$

which now holds by the Cauchy–Schwartz inequality.

**Appendix A 3.5.1**

**Proof of Lemma 3.5.1.** Let $\mathbf{B}_0 = \sum \phi_i \alpha_i \alpha_i'$ be the spectral decomposition of $\mathbf{B}_0$. Since $\mathbf{f}(L)$, $\mathbf{f}(s)$ and $\mathbf{f}(U)$ constitute a basis for $\mathbf{R}^3$, we can write $\mathbf{B}_0$ as

$$\begin{aligned} \mathbf{B}_0 &= d_{(LL)}\mathbf{f}(L)\mathbf{f}'(L) + d_{(ss)}\mathbf{f}(s)\mathbf{f}'(s) + d_{(UU)}\mathbf{f}(U)\mathbf{f}'(U) + d_{(Ls)}\mathbf{f}(L)\mathbf{f}'(s) + \\ &\quad \ldots + d_{(sU)}\mathbf{f}(s)\mathbf{f}'(U). \end{aligned} \tag{3.5.29}$$

We now substitute the above expression for $\mathbf{B}_0$ in $\mathrm{tr}\,\mathbf{M}^{-1}\mathbf{B}_0$ and use the expressions for $\mathbf{M}^{-1}\mathbf{f}(L)$, $\mathbf{M}^{-1}\mathbf{f}(s)$ and $\mathbf{M}^{-1}\mathbf{f}(U)$, given immediately after (3.5.23). The term involving $d_{(Ls)}$ is seen to be

$$d_{(Ls)}\,\mathrm{tr}[\gamma(L)\mathbf{f}'(s)/p\mathbf{f}'(L)\gamma(L)] = 0$$

since

$$\mathrm{tr}[\gamma(L)\mathbf{f}'(s)] = [\mathbf{f}'(s)\gamma(L)] = 0.$$

On the other hand, the term involving $d_{(LL)}$ is

$$d_{(LL)}\,\mathrm{tr}[\gamma(L)\mathbf{f}'(L)]/[p\mathbf{f}'(L)\gamma(L)] = \frac{d_{(LL)}}{p}.$$

Thus we see that

$$tr[\mathbf{M}^{-1}\mathbf{B}_0] = [\frac{d_{(LL)}}{p}] + [\frac{d_{(ss)}}{q}] + [\frac{d_{(UU)}}{r}].$$

To evaluate $d_{(LL)}$, we pre-multiply the above expression for $\mathbf{B}_0$ by $\gamma'(L)$ and post-multiply it by $\gamma(L)$. This yields

$$\gamma'(L)\mathbf{B}_0\gamma(L) = d_{(LL)}\gamma'(L)\mathbf{f}(L)\mathbf{f}'(L)\gamma(L)$$

and hence

$$d_{(LL)} = [\gamma'(L)\mathbf{B}_0\gamma(L)]/[\mathbf{f}'(L)\gamma(L)]^2.$$

Similarly we will have analogous expressions for $d_{(ss)}$ and $d_{(UU)}$. This completes the proof.

# References


Atkinson, A. C. and Donev, A. N. (1992). *Optimum experimental design.* Oxford, Oxford University Press.

de la Garza, A. (1954). Spacing of information in polynomial estimation. *Annals of Mathematical Statistics* **25**, 123–130.

Kiefer, J. C. (1959). Optimum experimental designs (with discussion). *Journal of the Royal Statistical Socciety* Series B **21**, 272–319.

Luoma, A., Mandal, N. K. and Sinha, Bikas K. (2001a). A-optimal cubic and quartic regression designs in asymmetric factor spaces. Submitted to *Statistics and Applications*. New Delhi.

Luoma, A., Nummi, T. and Sinha, Bikas K. (2001b): Optimal designs in random coefficient cubic regression models. Technical Report A 337, University of Tampere, Finland.

Liski, E. P., Luoma, A., Mandal, N. K. and Sinha, Bikas K. (1997). Optimal design for an inverse prediction problem under random coefficient regression models. *Journal of the Indian Society of Agricultural Statistics* Vol. XLIX Golden Jubilee Number 1996–1997, 277–288.

Liski, E. P., Luoma, A., Mandal, N. K. and Sinha, Bikas K. (1998). Optimal designs for prediction in random coefficient linear regression models. *Journal of Combinatorics, Information and System Sciences* (J. N. Srivastava Felicitation Volume), **23**(1–4), 1–16.

Liski, E. P., Luoma, A. and Sinha, Bikas K. (1996). Optimal designs in a random coefficient linear growth curve model. *Calcutta Statistical Association Bulletin* **46**, 211–229.

Mandal, N. K., Shah, K. R. and Sinha, Bikas K. (2000). de la Garza phenomenon re-visited. Unpublished Manuscript.

Pázman, A. (1986). *Foundations of optimum experimental design*. Reidel, Dordrecht.

Pukelsheim, F. (1993). *Optimal design of experiments*. Wiley, New York.

# 4

# Optimal Designs for Covariates' Models with Structured Intercept Parameter

## Summary

### Features

**Model(s):** Discrete design models with controllable covariates or multifactor linear regression models with structured intercept parameter
**Experimental domains:** Binary set-up for discrete designs and $\mathcal{T} = [-1, 1]$ for covariates
**Optimality criteria:** Most efficient estimation of treatment parameters and covariate parameters
**Major tools:** Mutually orthogonal latin squares and Hadamard matrices
**Optimality results:** Optimal designs under CRD, RBD and BIBD set-up
**Thrust:** Combinatorial optimization in terms of orthogonality (i) between factors versus blocks and treatments and also (ii) within the factors with levels $\pm 1$

This Chapter addresses optimality issues in a non-standard set-up. We envisage an experimental design situation where in every experimental unit, some controllable covariates are measured along with the response to the treatment applied. We present a variety of optimal designs in such a situation under different design settings. Optimality refers to the most efficient estimation of treatment parameters and covariate parameters simultaneously. We also discuss a related challenging issue viz., accommodating maximum number of covariates in the model without sacrificing the optimal nature of the underlying designs.

## 4.1 Introduction

In Chapters 2 and 3, we have examined in details the nature of regression designs under homoscedastic error structure - in the light of the de la Garza phenomenon. Also in Chapter 3, we have covered the cases of fixed and/or random regression coeffients in case of quadratic regression when our interest is in prediction or inverse prediction and the experimental domain is asymmetric in nature. In this Chapter, we consider models (in experimental design set-up) where treatments are compared with or without blocking in the presence of non-stochastic controllable covariates. In addition to the comparison of treatments, we are also interested in accommodating as many covariates as possible, subject to these being optimally estimated. Situations where the covariates are *not* under the control of the experimenter were discussed by Haggstrom (1975), Harville (1975) and Wu (1981). These are also briefly discussed in Shah and Sinha (1989).

Situations where the covariates are under the control of the experimenter were first considered by Lopes Troya (1982a, 1982b). This is the subject matter of the present Chapter. In Section 4.2 we discuss the situation without blocking whereas the situation with blocking is considered in Section 4.3. The construction of optimal designs requires use of *mutually orthogonal latin squares (MOLS)* as well as *Hadamard matrices.* This area is relatively unexplored and there is room for much further work here. This Chapter is largely based on the work of Das *et al.* (2000).

Traditionally, in a study of linear regression designs involving non-stochastic regressors, we tacitly call for "homogeneous" experimental units so that the assumed model for the $N \times 1$ observation vector $\mathbf{Y}$ is of the form

$$(\mathbf{Y}, \mu\mathbf{1} + \mathbf{Z}\gamma, \sigma^2\mathbf{I}), \tag{4.1.1}$$

where $\mu$ represents the intercept term, $\mathbf{1}$ denotes the vector of all ones, $\gamma$ is the vector of covariates' effects and $\mathbf{Z}$ is the matrix of covariates. Understandably, homogeneous nature of the experimental units *safe-guards the same intercept term* for every expectation of the observations in (4.1.1).

It is well known that when the experimental domain of the regressors is a $c$-dimensional cube of the form: $[-1, 1]^c$, the most efficient design for estimation of the regression coefficients (i.e., the $\gamma$-parameters) is derived from a Hadamard matrix, wherever the latter exists. For $N > c$, it is enough to start with a Hadamard matrix $\mathbf{H}_N$ of order $N$ (in its standard form) and select any $c$ of its columns for the matrix $\mathbf{Z}$, leaving the first column. This yields an optimum design for the (joint) estimation of $\mu$ and $\gamma$, on the basis of $N$ observations. Optimality here refers to attaining the least possible value $(\sigma^2/N)$ of the individual variance components simultaneously for all the parameter estimates.

The covariates' model studied by Lopes Troya stipulates that the observation vector $\mathbf{Y}$ has the model

$$(\mathbf{Y}, \mathbf{X}\tau + \mathbf{Z}\gamma, \sigma^2\mathbf{I}), \tag{4.1.2}$$

where in addition to $\gamma$-parameters, $\boldsymbol{\tau}$ is the vector of a set of (possibly different) $v$ treatment effects and $\mathbf{X}$ is the *incidence matrix* of the treatments among the experimental units.

This is referred to as "one-way ANOVA model (without the general mean) with covariates". We may perfectly interpret (4.1.2) as a model for the study of multivariate linear regression in situations where the common intercept term has been replaced by one depending on one "blocking factor", viz., the treatments effects. We may as well stipulate that the model (4.1.2) envisages a study of the treatments effects in a completely randomized design (CRD) set-up in the presence of the covariates.

Lopes Troya studied the nature of *optimal allocation of treatments and covariates* in the above set-up for simultaneous estimation of the (fixed) treatment effects (in the absence of the general effect) and the covariates effects with maximum efficiency in the sense of minimum generalized variance.

In this Chapter, we will undertake a detailed study of the above problem in a variety of situations:

(1) One-way ANOVA model with covariates,

(2) Two-way ANOVA model of the form

$$(\mathbf{Y}, \mu\mathbf{1} + \mathbf{X}_1\boldsymbol{\beta} + \mathbf{X}_2\boldsymbol{\tau} + \mathbf{Z}\boldsymbol{\gamma}, \sigma^2\mathbf{I}), \tag{4.1.3}$$

where in addition to $\mu$, $\gamma$ and $\boldsymbol{\tau}$, $\boldsymbol{\beta}$ represents the vector of (possibly different) block effects, $\mathbf{X}_1$ and $\mathbf{X}_2$ being the incidence matrices of block effects and treatment effects, respectively.

In all such situations, it turns out that, in effect, we are studying the nature of error functions in a linear model without covariates with emphasis on their structures for optimal estimation of covariates effects, in case these are present. Das *et al.* (2000) emphasize this aspect of data analysis and this seems to be an interesting area for further research.

## 4.2 Optimal Regression Designs with One-Way Classified Intercepts

Referring to the model (4.1.2), which excludes the general mean, the parameter vector of interest is given by

$$\eta = \begin{pmatrix} \tau \\ \gamma \end{pmatrix}. \tag{4.2.1}$$

For a given allocation vector $\mathbf{n} = (n_1, n_2, \ldots, n_v)'$ with a total of $N = \sum n_i$ observations, the information matrix for $\eta$ is given by

$$\mathcal{I} = \begin{pmatrix} \mathbf{X}'\mathbf{X} & \mathbf{X}'\mathbf{Z} \\ \mathbf{Z}'\mathbf{X} & \mathbf{Z}'\mathbf{Z} \end{pmatrix}. \tag{4.2.2}$$

The problem is to suggest an optimal allocation scheme (for given design parameters $N$, $v$ and $c$) in the presence of the covariates for efficient estimation of the treatment effects as well as the covariates effects. The choice of the levels $z_{ij}$ of the covariates is left to the experimenter and each level can vary within a given experimental domain, conveniently taken as $[-1,\ 1]$. This may be realized as a location-scale transformed version of the original arbitrary but finite limits.

It is evident that orthogonal estimation of treatment effects and covariate effects is possible whenever the orthogonality condition

$$\mathbf{Z}'\mathbf{X} = \mathbf{0} \tag{4.2.3}$$

is satisfied. Further, most efficient estimation of $\gamma$-components is possible whenever, in addition to (4.2.3), we can also ascertain the condition

$$\mathbf{Z}'\mathbf{Z} = N\mathbf{I}_c. \tag{4.2.4}$$

**Remark 4.2.1** It may be noted that whenever (4.2.3) is ensured, presence of the covariates in the model (4.1.2) does not pose any threat to the usual "optimal treatment allocation" problem. We divide $N$ into equal or almost equal allocation numbers so that $|n_i - n_j| \leq 1$ for all $i \neq j$.

Following Lopes Troya, we intend to discuss the availability of $\mathbf{Z}$ matrices satisfying (4.2.3) and (4.2.4) when the treatment allocation matrix $\mathbf{X}$ corresponds to an equal allocation number, i.e. in situations where $N$ is a multiple of $v$. We will write $N = vR$ so that $R$ is the common allocation number of the $v$ treatments under investigation. Note that there is also a related problem and this is what makes the whole issue *combinatorially challenging.* We must look out for the $\mathbf{Z}$ *matrix which involves the maximum number of parameters and yet satisfies* (4.2.3) *and* (4.2.4).

Not to obscure the essential steps of reasoning for this simple set-up as also for the block design set-up (to be taken up in the next section), let us have a representation of any column of the $\mathbf{Z}$-matrix (which is a column vector of order $N \times 1$) in the form of a matrix $\mathbf{W}$ of order $R \times v$:

| | Treatments: $1 \quad 2 \quad 3 \quad \ldots \quad v-1 \quad v$ | |
|---|---|---|
| $\mathbf{W}_{R\times v} =$ | $(\pm 1)_{R\times v}$ | |
| Column totals | $0 \quad 0 \quad 0 \quad \ldots \quad 0 \quad 0$ | $0$ = Total |
| | Essential conditions | A superfluous condition |

(4.2.5)

The columns of $\mathbf{W}$ correspond to the treatments in the order $1, 2, \ldots, v$, with the stipulation that $\mathbf{Z}$ satisfies both (4.2.3) and (4.2.4).

Conditions (4.2.3) and (4.2.4) imply that the $\mathbf{W}$-matrices satisfy the following conditions:

(1) The entries of all such **W**-matrices are $\pm 1$ and column totals are 0 each.

(2) For the Hadamard product of any pair of such matrices $\mathbf{W}^{(s)}$ and $\mathbf{W}^{(t)}$ defined as $\mathbf{W}^{(s)}_{R\times v} * \mathbf{W}^{(t)}_{R\times v} = \left(w^{(s)}_{ij} w^{(t)}_{ij}\right)_{R\times v}$, the total sum of all the elements is 0. In other words, $\sum_{i=1}^{R}\sum_{j=1}^{v} w^{(s)}_{ij} w^{(t)}_{ij} = 0$.

This is exemplified in the representation below:

$$
\begin{array}{l|cccccc|l}
 & \multicolumn{6}{c|}{\text{Treatments}} & \\
 & 1 & 2 & 3 & \ldots & v-1 & v & \\
\hline
\mathbf{W}^{(s)}_{R\times v} * \mathbf{W}^{(t)}_{R\times v} = & \multicolumn{6}{c|}{\left(w^{(s)}_{ij} w^{(t)}_{ij}\right)_{R\times v}} & \\
\hline
\text{Column totals} & 0 & 0 & 0 & \ldots & 0 & 0 & 0 = \sum_{i=1}^{R}\sum_{j=1}^{v} w^{(s)}_{ij} w^{(t)}_{ij} \\
 & \multicolumn{6}{c|}{\text{Desirable conditions}} & \text{An essential condition}
\end{array}
\tag{4.2.6}
$$

**Remark 4.2.2** In the matrix representations (4.2.5) and (4.2.6) above we have displayed essential, superfluous and desirable conditions to be satisfied by the elements of the **W**-matrices under consideration. This, in our estimation, brings out quite nicely the inherent nature of the combinatorial problems.

Matrices satisfying the desirable conditions in (4.2.6) will be called *very regular*; otherwise they will be termed as *regular*, so that the essential conditions are preserved. It turns out that whenever $R$ is an even integer, the condition (4.2.5) is trivially satisfied by choosing the elements of each column of **W** as $\pm 1$ in equal numbers. This is true for each such matrix taken separately. For $v = 2$, we may start with a $2 \times 2$ matrix $\mathbf{M}_{2\times 2}$ defined as

$$\mathbf{M}_{2\times 2} = \begin{pmatrix} 1 & -1 \\ -1 & 1 \end{pmatrix}.$$

Then we use $R/2$ repetitions of **M** to generate $\mathbf{W}^{(1)}$. For $\mathbf{W}^{(2)}$, we may retain the first column of $\mathbf{W}^{(1)}$ and change the signs of all the elements of the second column of $\mathbf{W}^{(1)}$. Thus $\mathbf{W}^{(2)}$ may be derived from $\mathbf{W}^{(1)}$ by the above rule.

Below we only treat the cases $R = 0 \pmod 4$, unless otherwise stated. We now present some results from Lopes Troya(1982a) with reference to the **W**-matrices in (4.2.5). In certain cases, these are *very regular* while in others, these are just *regular*. For Theorem 4.2.1(a)–(c), we will assume the existence of $\mathbf{H}_R$, the Hadamard matrix of order $R$. Also we will denote by c* the *maximum* value of c (the number of covariates) in a given context as attained by a given method of construction.

**Theorem 4.2.1** *There exist* **W**-*matrices satisfying*

(a) $c^* = R - 1$ *for* $v$ *odd,*

(b) $c^* = 2(R-1)$ *for* $v = 2 \pmod 4$,

(c) $c^* = 4(R-1)$ *for* $v = 0 \pmod 4$,

(d) $c^* = 3v$ *whenever* $\mathbf{H}_v$ *exists,*

(e) $c^* = v$ *whenever* $\mathbf{H}_v$ *exists and* $R = 2 \pmod 4$.

**Proof.** Let us represent a Hadamard matrix $\mathbf{H}_R$ of order $R$ as

$$\mathbf{H}_R = (\mathbf{h}_1,\ \mathbf{h}_2,\ \ldots, \mathbf{h}_{R-1},\ \mathbf{1}), \tag{4.2.7}$$

where the columns $\mathbf{h}_i$, $i = 1, 2, \ldots, R-1$ are $R \times 1$-vectors with entries $\pm 1$ and $\mathbf{1}$ is the $R \times 1$-vector of all ones. The choice of the **W**-matrices is indicated below one by one. The verification is immediate from (4.2.5) and (4.2.6) and we leave it to the readers.

Cases (a)–(c) correspond to *very regular* features while (d)–(e) relate to *regular* features. Case (d) is interesting in itself since it does *not* need existence of $\mathbf{H}_R$. Also for $R = 0 \pmod 4$ and $v = 0 \pmod 4$, it follows from (c) and (d) that $c^* = \max\{4(R-1), 3v\}$. See Remark 4.2.3 for a strengthened version of this result. In the sequel we denote $\tilde{\mathbf{h}}_j = -\mathbf{h}_j,\ 1 \le j \le R-1$. Then the choice of these **W**-matrices is as follows:

(a)

$$\mathbf{W}^{(j)}_{R\times v} = \underbrace{(\mathbf{h}_j, \mathbf{h}_j, \ldots, \mathbf{h}_j)}_{\leftarrow v \text{ columns} \rightarrow}, \quad 1 \le j \le R-1; \tag{4.2.8}$$

(b)

$$\mathbf{W}^{(j1)}_{R\times v} = (\underbrace{\mathbf{h}_j, \mathbf{h}_j, \ldots, \mathbf{h}_j}_{\leftarrow \frac{v}{2} \text{ columns} \rightarrow} \mid \underbrace{\mathbf{h}_j, \mathbf{h}_j, \ldots, \mathbf{h}_j}_{\leftarrow \frac{v}{2} \text{ columns} \rightarrow}), \quad 1 \le j \le R-1;$$

$$\mathbf{W}^{(j2)}_{R\times v} = (\underbrace{\mathbf{h}_j, \mathbf{h}_j, \ldots, \mathbf{h}_j}_{\leftarrow \frac{v}{2} \text{ columns} \rightarrow} \mid \underbrace{\tilde{\mathbf{h}}_j, \tilde{\mathbf{h}}_j, \ldots, \tilde{\mathbf{h}}_j}_{\leftarrow \frac{v}{2} \text{ columns} \rightarrow}), \quad 1 \le j \le R-1; \tag{4.2.9}$$

(c)

$$\mathbf{W}^{(j1)}_{R\times v} =$$

$$(\underbrace{\mathbf{h}_j, \mathbf{h}_j, \ldots, \mathbf{h}_j}_{\leftarrow \frac{v}{4} \text{ columns} \rightarrow} \mid \underbrace{\mathbf{h}_j, \mathbf{h}_j, \ldots, \mathbf{h}_j}_{\leftarrow \frac{v}{4} \text{ columns} \rightarrow} \mid \underbrace{\tilde{\mathbf{h}}_j, \tilde{\mathbf{h}}_j, \ldots, \tilde{\mathbf{h}}_j}_{\leftarrow \frac{v}{4} \text{ columns} \rightarrow} \mid \underbrace{\tilde{\mathbf{h}}_j, \tilde{\mathbf{h}}_j, \ldots, \tilde{\mathbf{h}}_j}_{\leftarrow \frac{v}{4} \text{ columns} \rightarrow})$$

$$\mathbf{W}^{(j2)}_{R\times v} =$$

$$(\mathbf{h}_j, \mathbf{h}_j, \ldots, \mathbf{h}_j \mid \tilde{\mathbf{h}}_j, \tilde{\mathbf{h}}_j, \ldots, \tilde{\mathbf{h}}_j \mid \mathbf{h}_j, \mathbf{h}_j, \ldots, \mathbf{h}_j \mid \tilde{\mathbf{h}}_j, \tilde{\mathbf{h}}_j, \ldots, \tilde{\mathbf{h}}_j)$$
$$\leftarrow \tfrac{v}{4} \text{ columns} \rightarrow \; \leftarrow \tfrac{v}{4} \text{ columns} \rightarrow \; \leftarrow \tfrac{v}{4} \text{ columns} \rightarrow \; \leftarrow \tfrac{v}{4} \text{ columns} \rightarrow$$

$$\mathbf{W}^{(j3)}_{R\times v} =$$
$$(\mathbf{h}_j, \mathbf{h}_j, \ldots, \mathbf{h}_j \mid \tilde{\mathbf{h}}_j, \tilde{\mathbf{h}}_j, \ldots, \tilde{\mathbf{h}}_j \mid \tilde{\mathbf{h}}_j, \tilde{\mathbf{h}}_j, \ldots, \tilde{\mathbf{h}}_j \mid \mathbf{h}_j, \mathbf{h}_j, \ldots, \mathbf{h}_j)$$
$$\leftarrow \tfrac{v}{4} \text{ columns} \rightarrow \; \leftarrow \tfrac{v}{4} \text{ columns} \rightarrow \; \leftarrow \tfrac{v}{4} \text{ columns} \rightarrow \; \leftarrow \tfrac{v}{4} \text{ columns} \rightarrow$$

$$\mathbf{W}^{(j4)}_{R\times v} =$$
$$(\mathbf{h}_j, \mathbf{h}_j, \ldots, \mathbf{h}_j \mid \mathbf{h}_j, \mathbf{h}_j, \ldots, \mathbf{h}_j \mid \mathbf{h}_j, \mathbf{h}_j, \ldots, \mathbf{h}_j \mid \mathbf{h}_j, \mathbf{h}_j, \ldots, \mathbf{h}_j)$$
$$\leftarrow \tfrac{v}{4} \text{ columns} \rightarrow \; \leftarrow \tfrac{v}{4} \text{ columns} \rightarrow \; \leftarrow \tfrac{v}{4} \text{ columns} \rightarrow \; \leftarrow \tfrac{v}{4} \text{ columns} \rightarrow$$
$$1 \le j \le R-1; \tag{4.2.10}$$

(d) It is enough to display the **W**-matrices for the case of $R = 4$. For higher multiples of 4 we may simply use identical copies.

The case $R = 4$ is resolved by using the following forms of **W**-matrices (shown as transposes): *Note that here we are using the same representation for* $\mathbf{H}_v = (\mathbf{h}_1, \mathbf{h}_2, \ldots, \mathbf{h}_{v-1}, \mathbf{1})$ *as before.* The **W**-matrices are as follows:

$$\begin{aligned}
\left[\mathbf{W}^{(j1)}_{4\times v}\right]' &= (\mathbf{h}_j, \mathbf{h}_j, \tilde{\mathbf{h}}_j, \tilde{\mathbf{h}}_j), \\
\left[\mathbf{W}^{(j2)}_{4\times v}\right]' &= (\mathbf{h}_j, \tilde{\mathbf{h}}_j, \mathbf{h}_j, \tilde{\mathbf{h}}_j), \\
\left[\mathbf{W}^{(j3)}_{4\times v}\right]' &= (\mathbf{h}_j, \tilde{\mathbf{h}}_j, \tilde{\mathbf{h}}_j, \mathbf{h}_j), \quad j = 1, 2, \ldots, v. \qquad (4.2.11)
\end{aligned}$$

(e) Here again, it is enough to settle the case of $R = 2$ which is done using the same technique as above and using the same representation for $\mathbf{H}_v$. The **W**-matrix is

$$\left[\mathbf{W}^{(j)}_{2\times v}\right]' = (\mathbf{h}_j, \tilde{\mathbf{h}}_j), \quad 1 \le j \le v.$$

□

**Remark 4.2.3** The result in Theorem 4.2.1(c) [(d)] is based on the tacit assumption that we only have available $\mathbf{H}_R$ (respectively $\mathbf{H}_v$) and not quite $\mathbf{H}_v$ (respectively $\mathbf{H}_R$). However, prompted by the condition $v = R = 0 \pmod 4$, we can increase the value of $c^*$ to $c^* = v(R-1)$ by assuming existence of both $\mathbf{H}_v$ and $\mathbf{H}_R$. To see this, we can argue as follows:

Set

$$\mathbf{H}_R = (\mathbf{h}_1, \mathbf{h}_2, \ldots, \mathbf{h}_{R-1}, \mathbf{1}) \text{ and } \mathbf{H}'_v = (\mathbf{h}^*_1, \mathbf{h}^*_2, \ldots, \mathbf{h}^*_{v-1}, \mathbf{1})$$

and construct $\mathbf{W}^{(ij)}$ by choosing

$$\mathbf{W}^{(ij)} = \mathbf{h}^{*'}_i \otimes \mathbf{h}_j = \left(w^{(ij)}_{rs}\right);$$

$r = 1, 2, \ldots, R$; $s = 1, 2, \ldots, v$; $i = 1, 2, \ldots, R-1$; $j = 1, 2, \ldots, v$. Here "$\otimes$" refers to the Kronecker product (vide Rao 1973).

It can be readily verified that

(i) $\mathbf{1}'(\mathbf{W}^{(ij)} * \mathbf{W}^{(ij')})\mathbf{1} = 0$ since $\sum_r w_{rs}^{(ij)} w_{rs}^{(ij')} = 0$ for every $s$,

(ii) $\mathbf{1}'(\mathbf{W}^{(ij)} * \mathbf{W}^{(i'j)})\mathbf{1} = 0$ since $\sum_s w_{rs}^{(ij)} w_{rs}^{(i'j)} = 0$ for every $r$,

(iii) $\mathbf{1}'(\mathbf{W}^{(ij)} * \mathbf{W}^{(i'j')})\mathbf{1} = 0$ since $\sum_r w_{rs}^{(ij)} w_{rs}^{(i'j')} = 0$ for every $s$.

For example, with $v = R = 4$, we can accommodate 12 **W**-matrices. These are given by

$$\begin{array}{llllll}
\mathbf{W}^{(11)} & = \mathbf{h}_1^{*'} \otimes \mathbf{h}_1 & \mathbf{W}^{(21)} & = \mathbf{h}_2^{*'} \otimes \mathbf{h}_1 & \mathbf{W}^{(31)} & = \mathbf{h}_3^{*'} \otimes \mathbf{h}_1 \\
\mathbf{W}^{(12)} & = \mathbf{h}_1^{*'} \otimes \mathbf{h}_2 & \mathbf{W}^{(22)} & = \mathbf{h}_2^{*'} \otimes \mathbf{h}_2 & \mathbf{W}^{(31)} & = \mathbf{h}_3^{*'} \otimes \mathbf{h}_2 \\
\mathbf{W}^{(13)} & = \mathbf{h}_1^{*'} \otimes \mathbf{h}_3 & \mathbf{W}^{(23)} & = \mathbf{h}_2^{*'} \otimes \mathbf{h}_3 & \mathbf{W}^{(33)} & = \mathbf{h}_3^{*'} \otimes \mathbf{h}_3 \\
\mathbf{W}^{(14)} & = \mathbf{h}_1^{*'} \otimes \mathbf{h}_4 & \mathbf{W}^{(24)} & = \mathbf{h}_2^{*'} \otimes \mathbf{h}_4 & \mathbf{W}^{(34)} & = \mathbf{h}_3^{*'} \otimes \mathbf{h}_4,
\end{array}$$

where

$$\mathbf{H}_4 = \begin{pmatrix} -1 & -1 & 1 & 1 \\ 1 & -1 & -1 & 1 \\ -1 & 1 & -1 & 1 \\ 1 & 1 & 1 & 1 \end{pmatrix} = (\mathbf{h}_1, \mathbf{h}_2, \mathbf{h}_3, \mathbf{1}) = \begin{pmatrix} \mathbf{h}_1^{*'} \\ \mathbf{h}_2^{*'} \\ \mathbf{h}_3^{*'} \\ \mathbf{1}' \end{pmatrix}$$

is a Hadamard matrix of order 4 and $\mathbf{h}_i$, $\mathbf{h}_i^*$; $i = 1, 2, 3, 4$, are $4 \times 1$ column vectors.

Finally, note that in the process, the value of $c^*$ has reached its absolute upper bound since in a CRD involving $v$ treatments each with replication $R$ and without any covariates, the error degrees of freedom is $v(R-1)$. Lopes Troya (1982a) may not have paid much attention to this point.

**Remark 4.2.4** In Lopes Troya (1982a,b) both regular and non-regular cases have been extensively discussed under one-way ANOVA set-up (in the absence of the general mean). Here non-regular cases correspond to the situations where (4.2.3) and/or (4.2.4) are violated, and yet, maximum possible efficiency for estimation of $\gamma$-components is achieved. We briefly study some features of the non-regular case below in a model without the general mean. In regular cases, it turns out that the model with the general mean does not pose any additional problem.

We examine the situations where the symmetrical allocation (SA) does *not* exist. That is, we consider situations wherein $N = vn_0 + p$, $0 < p < v$. In the absence of the covariates, it is known that the most symmetrical allocation (MSA) viz., one for which $|n_i - n_j| \leq 1, 1 \leq i < j \leq v$ is A-, D- and E-optimal for inference on the treatment means.

We intend to examine the availability of *most efficient* regression designs for estimation of the $\gamma$-parameters, using the above MSA. We now refer to a model *without* the general mean. From the set-up, it is clear that we can not simultaneously attain (4.2.3) and (4.2.4), based on the MSA, even for $c = 1$. Note that the information matrix for the $\gamma$-parameters is given by

$$\begin{aligned} \mathcal{I}(\gamma) &= \mathbf{Z}'\mathbf{Z} - \mathbf{Z}'\mathbf{X}(\mathbf{X}'\mathbf{X})^{-1}\mathbf{X}'\mathbf{Z} \\ &= \left( \sum_{i=1}^{v} \sum_{j=1}^{n_i} \left( z_{ij}^{(s)} - \bar{z}_i^{(s)} \right) \left( z_{ij}^{(s')} - \bar{z}_i^{(s')} \right) \right)_{1 \le s, s' \le c}, \end{aligned} \tag{4.2.12}$$

where $\mathbf{X}'\mathbf{X} = \text{Diag}(n_1, n_2, \ldots, n_v)$. Therefore, the best possible choice of $\mathbf{Z}'\mathbf{X}$ should be a matrix whose all entries are 0 or $\pm 1$.

This can be seen as follows: From (4.2.12), it follows that individual diagonal elements of $\mathcal{I}(\gamma)$ will be marginally maximized whenever

$$z_{ij}^{(s)} \in \{0, \pm 1\} \text{ for all } \; i, j, s \tag{4.2.13}$$

subjcct to

$$\bar{z}_i^{(s)} \in \{0, \pm 1/n_i\} \text{ for all } i, \text{ and } s, \tag{4.2.14}$$

where $\bar{z}_i^{(s)} = u_i^{(s)}/n_i$ for all $i$ and $s$. This along with

$$\sum_i \sum_j z_{ij}^{(s)} z_{ij}^{(s')} = 0 \text{ for all } s \neq s' \tag{4.2.15}$$

would result into an information matrix with diagonal and off-diagonal elements given by

$$\mathcal{I}(\gamma) = \left( N - \sum_i (u_i^{(s)})^2/n_i, - \sum_i u_i^{(s)} u_i^{(s')}/n_i \right), \tag{4.2.16}$$

where $u_i^{(s)} = 0, \pm 1$ for all $i$ and $s$.

This in a way leads to the *most efficient* allocation for the estimation of covariates parameters. Below we display the schematic version of the construction. Recall $\mathbf{W}$ has $N = vn_0 + p$ elements.

$$\mathbf{W} = \begin{array}{c} n_0 \left\{ \begin{array}{c} \\ \end{array} \right. \\ \\ \end{array} \left. \begin{array}{|c|c|} \hline (\pm 1) & (\pm 1) \\ \hline \text{Void} & \text{Extra row with } \pm 1 \\ \hline \longleftarrow v - p \longrightarrow & \leftarrow p \longrightarrow \end{array} \right\} n_0 + 1 \tag{4.2.17}$$

**Case 1:** $n_0 = 0 \pmod 4$ such that $\mathbf{H}_{n_0}$ exists.

**Case 2:** $n_0 + 1 = 0 \pmod 4$ such that $\mathbf{H}_{n_0+1}$ exists.

**Result for Case 1**: There exists a design with $c^* = \min\{n_0 - 1, p\}$ if $\mathbf{H}_{n_0}$ and $\mathbf{H}_p$ exist.

**Proof** (By construction.)

Write

$$\mathbf{H}_{n_0} = (\mathbf{h}_1, \mathbf{h}_2, \ldots, \mathbf{h}_{n_0-1}, \mathbf{1}), \quad \mathbf{H}_p^* = \begin{pmatrix} \mathbf{h}_1^{*'} \\ \mathbf{h}_2^{*'} \\ \vdots \\ \mathbf{h}_{p-1}^{*'} \\ \mathbf{1}' \end{pmatrix} \tag{4.2.18}$$

and

$$\begin{aligned} \mathbf{W}^{(s)} &= \left[\begin{array}{l|l} \mathbf{h}_s, \mathbf{h}_s, \ldots, \mathbf{h}_s & \mathbf{h}_s, \mathbf{h}_s, \ldots \mathbf{h}_s \\ \hline \text{Void} & \text{Extra row } \mathbf{h}_s^{*'} \end{array}\right] \begin{array}{l} \}\, n_0 \\ \}\, 1 \end{array} \\ s &= 1, 2, \ldots, c^*. \end{aligned}$$

The verification is now immediate.

**Result for Case 2**: There exists a design for which $c^* = \min\{n_0, v-p-1\}$ if $\mathbf{H}_{n_0+1}$ and $\mathbf{H}_{v-p}$ exist.

**Proof** (By construction.) As before, we write

$$\mathbf{H}_{n_0+1} = (\mathbf{h}_1, \mathbf{h}_2, \ldots, \mathbf{h}_{n_0}, \mathbf{1}) \text{ and } \mathbf{H}_{v-p}^* = \begin{pmatrix} \mathbf{h}_1^{*'} \\ \mathbf{h}_2^{*'} \\ \vdots \\ \mathbf{h}_{v-p-1}^{*'} \\ \mathbf{1}' \end{pmatrix}. \tag{4.2.19}$$

We construct

$$\mathbf{W}^{(s)} = \left[\begin{array}{l|l} \underline{\mathbf{h}}_1^{(s)}, \underline{\mathbf{h}}_2^{(s)}, \ldots, \underline{\mathbf{h}}_{v-p}^{(s)} & \mathbf{h}_s, \mathbf{h}_s, \ldots, \mathbf{h}_s \\ \hline \text{Void} & \end{array}\right], \tag{4.2.20}$$

where

$$\left(\mathbf{h}_1^{(s)}, \mathbf{h}_2^{(s)}, \ldots, \mathbf{h}_{v-p}^{(s)}\right) = (\mathbf{h}_s, \mathbf{h}_s, \ldots, \mathbf{h}_s) \otimes \mathbf{h}_s^{*'} \tag{4.2.21}$$

and $\underline{\mathbf{h}}_i^{(s)}$ $(i = 1, 2, \ldots, v-p)$ is $\mathbf{h}_i^{(s)}$ with the last element deleted. It is easy to verify that $\mathbf{W}^{(s)}$-matrices constructed above satisfy (4.2.14)–(4.2.15).

**Illustrative Examples:**

**Case 1:** $N = 52$, $v = 11$, $n_0 = 4$ and $p = 8$

$$\mathbf{W}^{(s)} = \begin{array}{|c|c|}\hline \mathbf{h}_s, \mathbf{h}_s, \mathbf{h}_s & \mathbf{h}_s, \mathbf{h}_s, \ldots, \mathbf{h}_s \\ \hline \text{Void} & \text{Extra row } = \mathbf{h}_s^{*'} \\ \hline \end{array},$$

$$\mathbf{H}_4 = (\mathbf{h}_1, \mathbf{h}_2, \mathbf{h}_3, \mathbf{1}) \text{ and } \mathbf{H}_8^* = \begin{pmatrix} \mathbf{h}_1^{*'} \\ \mathbf{h}_2^{*'} \\ \vdots \\ \mathbf{1}' \end{pmatrix} \quad \text{for } s = 1, 2, 3.$$

**Case 2:** $N = 68$, $v = 9$, $n_0 = 7$ and $p = 5$

$$\mathbf{W}^{(s)} = \begin{array}{|c|c|}\hline \underline{\underline{\mathbf{h}}}_1^{(s)}, \underline{\underline{\mathbf{h}}}_2^{(s)}, \underline{\underline{\mathbf{h}}}_3^{(s)}, \underline{\underline{\mathbf{h}}}_4^{(s)} & \mathbf{h}_s, \mathbf{h}_s, \ldots, \mathbf{h}_s \\ \cline{1-1} \text{Void} & \\ \hline \end{array},$$

where $\underline{\underline{\mathbf{h}}}_i^{(s)}$ is $\mathbf{h}_i^{(s)}$ with the last element deleted. Further,

$$\mathbf{H}_8 = (\mathbf{h}_1, \mathbf{h}_2, \ldots, \mathbf{h}_7, \mathbf{1})$$

and

$$\mathbf{H}_4^* = \begin{pmatrix} \mathbf{h}_1^{*'} \\ \mathbf{h}_2^{*'} \\ \mathbf{h}_3^{*'} \\ \mathbf{1}' \end{pmatrix} = \begin{pmatrix} 1 & -1 & 1 & -1 \\ 1 & 1 & -1 & -1 \\ 1 & -1 & -1 & 1 \\ 1 & 1 & 1 & 1 \end{pmatrix}.$$

For $s = 1$, 2 and 3 the arrays $\mathbf{W}^{(s)}$ are explicitly written as follows:

$$\mathbf{W}^{(1)} = \begin{array}{|c|c|}\hline \underline{\underline{\mathbf{h}}}_1, \tilde{\underline{\underline{\mathbf{h}}}}_1, \underline{\underline{\mathbf{h}}}_1, \tilde{\underline{\underline{\mathbf{h}}}}_1 & \mathbf{h}_1, \mathbf{h}_1, \mathbf{h}_1, \mathbf{h}_1, \mathbf{h}_1 \\ \cline{1-1} \text{Void} & \\ \hline \end{array}$$

$$\mathbf{W}^{(2)} = \begin{array}{|c|c|}\hline \underline{\underline{\mathbf{h}}}_2, \underline{\underline{\mathbf{h}}}_2, \tilde{\underline{\underline{\mathbf{h}}}}_2, \tilde{\underline{\underline{\mathbf{h}}}}_2 & \mathbf{h}_2, \mathbf{h}_2, \mathbf{h}_2, \mathbf{h}_2, \mathbf{h}_2 \\ \cline{1-1} \text{Void} & \\ \hline \end{array}$$

$$\mathbf{W}^{(3)} = \begin{array}{|c|c|}\hline \underline{\underline{\mathbf{h}}}_3, \tilde{\underline{\underline{\mathbf{h}}}}_3, \tilde{\underline{\underline{\mathbf{h}}}}_3, \underline{\underline{\mathbf{h}}}_3 & \mathbf{h}_3, \mathbf{h}_3, \mathbf{h}_3, \mathbf{h}_3, \mathbf{h}_3 \\ \cline{1-1} \text{Void} & \\ \hline \end{array}$$

Here $\tilde{\mathbf{h}}_i = -\mathbf{h}_i$ and $\tilde{\underline{\underline{\mathbf{h}}}}_i = -\underline{\underline{\mathbf{h}}}_i$, where $\underline{\underline{\mathbf{h}}}_i$ is $\mathbf{h}_i$ with the last element deleted.

## 4.3 Optimal Regression Designs with Two-Way Classified Intercepts

The key reference to this section is Das *et al.* (2000). This time we assume, for the response in $c$-factor linear regression set-up, the model

$$(\mathbf{Y}, \mu\mathbf{1} + \mathbf{X}_1\boldsymbol{\beta} + \mathbf{X}_2\boldsymbol{\tau} + \mathbf{Z}\boldsymbol{\gamma}, \sigma^2\mathbf{I}) \tag{4.3.1}$$

which corresponds to a situation where the intercept term is affected by two "blocking factors" i.e., the blocks and the treatments in a conventional block design.

We straightaway compute the form of the information matrix for the whole set of parameters underlying a design $d$ with $\mathbf{X}_{1d}$, $\mathbf{X}_{2d}$ and $\mathbf{Z}_d$ as the versions of $\mathbf{X}_1$, $\mathbf{X}_2$ and $\mathbf{Z}$ in (4.3.1):

$$\mathcal{I}(d) = \begin{pmatrix} N & \mathbf{1}'\,\mathbf{X}_{1d} & \mathbf{1}'\,\mathbf{X}_{2d} & \mathbf{1}'\,\mathbf{Z}_d \\ & \mathbf{X}'_{1d}\mathbf{X}_{1d} & \mathbf{X}'_{1d}\mathbf{X}_{2d} & \mathbf{X}'_{1d}\mathbf{Z}_d \\ & & \mathbf{X}'_{2d}\mathbf{X}_{2d} & \mathbf{X}'_{2d}\mathbf{Z}_d \\ & & & \mathbf{Z}'_d\mathbf{Z}_d \end{pmatrix}. \qquad (4.3.2)$$

For the covariates, we assume the (location-scale)-transformed version: $|z_{ij}| \leq 1$.

It is evident that orthogonal estimation of treatment and block effects contrasts on the one hand and covariates effect on the other is possible when the conditions

$$\mathbf{Z}'_d\mathbf{1} = \mathbf{0},\ \mathbf{Z}'_d\mathbf{X}_{1d} = \mathbf{0},\ \mathbf{Z}'_d\mathbf{X}_{2d} = \mathbf{0} \qquad (4.3.3)$$

are satisfied. Further, most efficient estimation of $\gamma$-components is possible whenever, in addition to (4.3.3), we can also ascertain the condition

$$\mathbf{Z}'_d\mathbf{Z}_d = N\mathbf{I}_c, \qquad (4.3.4)$$

$N$ being the order of $\mathbf{Y}$.

It is also true that whenever (4.3.3) is ensured, presence of the covariates in (4.3.1) does not pose any threat to the usual *optimal design* problem in a block design set-up.

For the rest of the chapter, we will mainly deal with a randomized block design (RBD) set-up and examine a variety of situations for existence and actual construction of designs satisfying (4.3.3) and (4.3.4) - with a special effort to accommodate maximum number of covariates ($c^*$). It turns out that, apart from the Hadamard matrices, the MOLS (of appropriate orders) are very useful in such endeavors.

### 4.3.1 Optimal Regression Designs in an RBD Set-Up

We have an RBD set-up with $b$ blocks and $v$ treatments so that $N = bv$. A matrix $\mathbf{Z} = (\pm 1)_{N\times c}$ will satisfy conditions (4.3.3) and (4.3.4) displayed above whenever, diagramatically, we can achieve (4.3.5) and (4.3.6) respectively. Recall (4.2.5) and (4.2.6) in this context.

$$
\begin{array}{r|c|l}
 & \text{Treatments} & \text{Row totals} \\
\hline
 & & 0 \\
 & & 0 \\
\text{Blocks} & \mathbf{W}^{(s)} = (\pm 1)_{b\times v} & \vdots \\
 & & 0 \\
\hline
\text{Column totals} & 0 \quad 0 \quad \cdots \quad 0 & 0
\end{array}
$$

Row totals: Essential conditions $1 \le s \le c$ (4.3.5)

Column totals: Essential conditions

Superfluous condition (Total)

$$
\begin{array}{r|c|l}
 & \text{Treatments} & \text{Row totals} \\
\hline
 & & 0 \\
 & & 0 \\
\text{Blocks} & \mathbf{W}^{(s)} * \mathbf{W}^{(t)} = (w_{ij}^{(s)} w_{ij}^{(t)}) & \vdots \\
 & & 0 \\
\hline
\text{Column totals} & 0 \quad 0 \quad \cdots \quad 0 & 0
\end{array}
$$

Row totals: Desirable conditions $1 \le s < t \le c$

Column totals: Desirable conditions

Essential condition (Total)

(4.3.6)

While providing constructions of **W**-matrices, we will mostly attempt to meet the desirable features and distinguish such cases by referring to them as *very regular* - as against the others which will be referred to as just *regular*. Recall that this is what was done also in connection with the CRD set-up.

### 4.3.2 Constructional Procedures when $b$ and $v$ Are Even Integers and $c^* = 1$

This case is easy to resolve. For the sake of future use, we present the following proposition, the proof of which is omitted. This shows that $c^* \ge 1$.

**Proposition 4.3.1** *Define* $\mathbf{M}_{2\times 2} = \begin{pmatrix} 1 & -1 \\ -1 & 1 \end{pmatrix}$. *Then for even integer $b$ and $v$*

$$\mathbf{W}_{b\times v} = \mathbf{J}_{b/2\times v/2} \otimes \mathbf{M}_{2\times 2} \tag{4.3.7}$$

*satisfies* (4.3.5).

### 4.3.3 Constructional Procedures for $b = v = 0 \pmod 4$

The following theorem brings out the full spirit of application of the MOLS of order $v$.

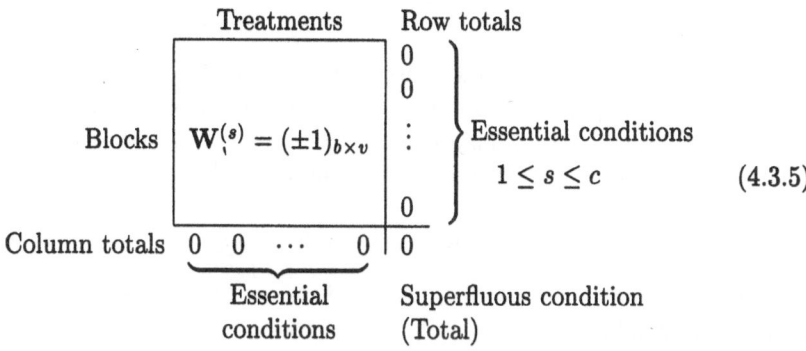

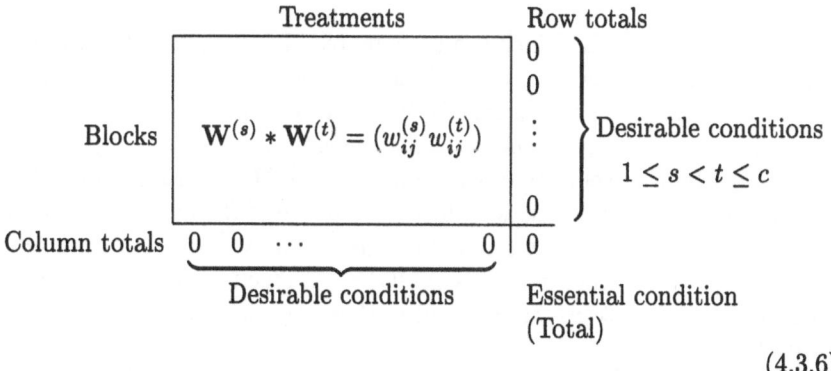

**Theorem 4.3.1** *Suppose $\mathbf{H}_v$ and further, m MOLS of order v exist. Then* $c^* = m(v-1)$.

**Proof.** We will provide actual construction of the **W**-matrices. Not to obscure the essential steps of reasoning, we will proceed as follows:

**Step 1:** We set the Hadamard matrix $\mathbf{H}_v$ in the form (watch out the difference between this representation and that immediately before (4.2.11))

$$\mathbf{H}_v = (\mathbf{h}_1, \mathbf{h}_2, \ldots, \mathbf{h}_{v-2}, \mathbf{h}_{v-1}, \mathbf{1})'. \tag{4.3.8}$$

**Step 2:** We form the $i$th member $\mathbf{L}_i$ of the set of m MOLS of order $v$ by using the symbols

$$A_{i1},\ A_{i2},\ \ldots,\ A_{iv};\ \ 1 \le i \le m \tag{4.3.9}$$

**Step 3:** Consider $\mathbf{L}_1$ which is a $v \times v$ matrix. In it, replace the symbols $A_{11}, A_{12}, \ldots, A_{1v}$ each occurring $v$ times, by the elements of the first row of $\mathbf{H}_v$. This will result into a matrix **W** to account for the first covariate parameter $\gamma_1$. Similarly, replacing the elements $A_{11}, A_{12}, \ldots, A_{1v}$ of $\mathbf{L}_1$ respectively by the elements of the second row of $\mathbf{H}_v$, we can account for the second covariate $\gamma_2$. This way, by exclusive use of the initial $v-1$ rows of $\mathbf{H}_v$ along with $\mathbf{L}_1$, we can accommodate as many as $v-1$ covariates. Likewise, use of the other members of MOLS will result in accommodations of covariates in sets of $(v-1)$. Thus altogether, we can account for $m(v-1)$ covariates.
That all such resulting components of the $\mathbf{Z}_{N\times c^*}$-matrix in (4.3.1) satisfy (4.3.5) and (4.3.6) are easy to verify. □

**Corollary 4.3.1** *When $b = v = 2^q$, q an integer, we have $c^* = (b-1)(v-1)$.*

**Proof.** For $v = 2^q$, we have a complete set i.e., $(v-1)$ MOLS of order $v$. The rest is clear. □

**Remark 4.3.1** It is evident that $c^*$ cannot exceed the above limit as it exhausts the error degrees of freedom in the RBD model.

**Corollary 4.3.2** *When $b = pv$, $v = 0 \pmod 4$, $p > 1$ and $\mathbf{H}_v$ and m MOLS of order v exist, then also we have $c^* = m(v-1)$.*

**Proof.** We work with the **W**-matrix in (4.3.5) as constructed in Theorem 3.2. This matrix has order $v \times v$. It is enough to repeat it $p$ times, one below the other, to form the resulting **W**-matrix of order $b \times v$. □

**Example 4.3.1** We will elaborate on the above method of construction by citing an example. We take $b = 4$, $v = 4$ and, for brevity, write down the MOLS of order 4 as follows:

$$\mathbf{L}_1 = \begin{pmatrix} a & b & c & d \\ b & a & d & c \\ c & d & a & b \\ d & c & b & a \end{pmatrix}, \quad \mathbf{L}_2 = \begin{pmatrix} \alpha & \delta & \beta & \gamma \\ \beta & \gamma & \alpha & \delta \\ \gamma & \beta & \delta & \alpha \\ \delta & \alpha & \gamma & \beta \end{pmatrix}, \quad \mathbf{L}_3 = \begin{pmatrix} p & s & r & q \\ q & r & s & p \\ s & p & q & r \\ r & q & p & s \end{pmatrix}.$$

The form of the $\mathbf{H}_4$ matrix for our use is:

$$\mathbf{H}_4 = \begin{pmatrix} 1 & -1 & 1 & -1 \\ 1 & 1 & -1 & -1 \\ 1 & -1 & -1 & 1 \\ 1 & 1 & 1 & 1 \end{pmatrix}.$$

Here are the $\mathbf{W}$-matrices accommodating the maximum possible number (9) of covariates in an optimal manner:

$$\begin{pmatrix} 1 & -1 & 1 & -1 \\ -1 & 1 & -1 & 1 \\ 1 & -1 & 1 & -1 \\ -1 & 1 & -1 & 1 \end{pmatrix} \begin{pmatrix} 1 & 1 & -1 & -1 \\ 1 & 1 & -1 & -1 \\ -1 & -1 & 1 & 1 \\ -1 & -1 & 1 & 1 \end{pmatrix} \begin{pmatrix} 1 & -1 & -1 & 1 \\ -1 & 1 & 1 & -1 \\ -1 & 1 & 1 & -1 \\ 1 & -1 & -1 & 1 \end{pmatrix}$$

$$\begin{pmatrix} 1 & -1 & 1 & -1 \\ 1 & -1 & 1 & -1 \\ -1 & 1 & -1 & 1 \\ -1 & 1 & -1 & 1 \end{pmatrix} \begin{pmatrix} 1 & 1 & -1 & -1 \\ -1 & -1 & 1 & 1 \\ -1 & -1 & 1 & 1 \\ 1 & 1 & -1 & -1 \end{pmatrix} \begin{pmatrix} 1 & -1 & -1 & 1 \\ -1 & 1 & 1 & -1 \\ 1 & -1 & -1 & 1 \\ -1 & 1 & 1 & -1 \end{pmatrix}$$

$$\begin{pmatrix} 1 & -1 & 1 & -1 \\ -1 & 1 & -1 & 1 \\ -1 & 1 & -1 & 1 \\ 1 & -1 & 1 & -1 \end{pmatrix} \begin{pmatrix} 1 & 1 & -1 & -1 \\ -1 & -1 & 1 & 1 \\ 1 & 1 & -1 & -1 \\ -1 & -1 & 1 & 1 \end{pmatrix} \begin{pmatrix} 1 & -1 & -1 & 1 \\ 1 & -1 & -1 & 1 \\ -1 & 1 & 1 & -1 \\ -1 & 1 & 1 & -1 \end{pmatrix}.$$

Next we aim at slightly strengthening the result of Theorem 4.3.1 in the sense of increasing the number of covariates from $m(v-1)$ to $m(v-1)+1$.

**Theorem 4.3.2** *Suppose $b$ is an even multiple of $v$ for which $\mathbf{H}_v$ exists. Then $c^* = m(v-1)+1$, where $m$ is the number of MOLS of order $v$.*

**Proof.** We may take $b = 2v$ without loss of generality. All the $\mathbf{W}$-matrices (of order $v \times v$) constructed earlier are retained and expanded by taking their negatives as well, in order to take care of $b = 2v$. An additional $\mathbf{W}$-matrix is given by $(\mathbf{J}_v, -\mathbf{J}_v)'$. □

We will now discuss some other isolated cases one by one.

### 4.3.4 Miscellaneous Results

**Result 4.3.1** *For* $b = 0 \pmod 4$ *and* $v = 0 \pmod 4$, *we can achieve* $c^* = 4$ *without any further condition.*

**Proof.** The proof is carried out by actual construction, as detailed below. We do first the case $b = v = 4$.

**Step 1:** Start with

$$\mathbf{Z}_1 = \begin{pmatrix} 1 & 1 & -1 & -1 \\ 1 & -1 & -1 & 1 \\ -1 & -1 & 1 & 1 \\ -1 & 1 & 1 & -1 \end{pmatrix}. \tag{4.3.10}$$

**Step 2:** Form $\mathbf{Z}_2$ by cyclical permutation of the columns of $\mathbf{Z}_1$.

**Step 3:** Form $\mathbf{Z}_3$ by cyclical permutation of the columns of $\mathbf{Z}_2$ and then convert it to $\mathbf{Z}_3^*$ by changing the signs of elements of columns 2 and 4.

**Step 4:** Form $\mathbf{Z}_4$ by cyclical permutation of the columns of $\mathbf{Z}_3$ and then convert it to $\mathbf{Z}_4^*$ by changing the signs of elements of columns 1 and 3. This can also be obtained just by interchanging columns 2 and 4 of $\mathbf{Z}_2$. The matrices $\mathbf{Z}_2$, $\mathbf{Z}_3^*$ and $\mathbf{Z}_4^*$ are displayed below for ready reference.

$$\mathbf{Z}_2 = \begin{pmatrix} -1 & 1 & 1 & -1 \\ 1 & 1 & -1 & -1 \\ 1 & -1 & -1 & 1 \\ -1 & -1 & 1 & 1 \end{pmatrix} \quad \mathbf{Z}_3^* = \begin{pmatrix} -1 & 1 & 1 & -1 \\ -1 & -1 & 1 & 1 \\ 1 & -1 & -1 & 1 \\ 1 & 1 & -1 & -1 \end{pmatrix}. \tag{4.3.11}$$

$$\mathbf{Z}_4^* = \begin{pmatrix} -1 & -1 & 1 & 1 \\ 1 & -1 & -1 & 1 \\ 1 & 1 & -1 & -1 \\ -1 & 1 & 1 & -1 \end{pmatrix}. \tag{4.3.12}$$

It is readily checked that these $\mathbf{Z}$-matrices serve as the $\mathbf{W}$-matrices in (4.3.5) and (4.3.6) in case of $b = v = 4$. For the general case, it is enough to augment these matrices along both directions. □

**Result 4.3.2** *For* $v = 4$ *and* $b = 2 \pmod 4$, *we can achieve* $c^* = 3$ *without any further condition.*

**Proof.** We first display the $\mathbf{W}_1$-matrix for the case of $b = 10$ and $v = 4$ in the form of *transpose of a block matrix*. The columns of $\mathbf{W}_1'$ constitute three separate block matrices below:

$$\mathbf{W}_1' = \left(\begin{array}{ccc|ccc|cccc} 1 & 1 & 1 & -1 & -1 & -1 & 1 & 1 & -1 & -1 \\ -1 & 1 & -1 & 1 & -1 & 1 & 1 & -1 & -1 & 1 \\ 1 & -1 & -1 & -1 & 1 & 1 & -1 & -1 & 1 & 1 \\ -1 & -1 & 1 & 1 & 1 & -1 & -1 & 1 & 1 & -1 \end{array}\right). \tag{4.3.13}$$

The first block matrix in (4.3.13) is derived from $\mathbf{H}_4$ (Example 4.3.1), by deleting its first (normalized) column. The second block matrix is its negative. The third block matrix is, in fact, the $\mathbf{Z}_1$ matrix in (4.3.10).

The other two $\mathbf{W}$-matrices are derived from $\mathbf{W}_1$ as follows:

$\mathbf{W}_2$-matrix is obtained from $\mathbf{W}_1$ by cyclical permutation of columns within each of the block matrices 1, 2 and 3. It may be noted that the permutation of block 3 ($\mathbf{Z}_1$) results in $\mathbf{Z}_2$ in (4.3.11). $\mathbf{W}_3$-matrix is obtained from $\mathbf{W}_2$ again by cyclical permutation of columns within each of the block matrices 1, 2 and 3 - along with necessary changes in the resulting $\mathbf{Z}_3$ -matrix to bring it to the form of $\mathbf{Z}_3^*$-matrix as in (4.3.11).

We now complete the rest of the argument as follows: For $b = 2 \pmod 4$, it is evident that $b$ is of the form: $b = 4(t-1)+6$. The first two block matrices in each of the $\mathbf{W}$-matrices above together take care of the order $4 \times 6$ while the last block is of the order $4 \times 4$. Hence for a given b of the form given above, we repeat the last block $(t-1)$ times. This completes the proof.

□

### 4.3.5 Optimum Choice of Covariates in a BIBD Set-Up

We now deal with the situation wherein the block design set-up admits existence of *a balanced incomplete block design (BIBD)* with the parameters $b$, $v$, $r$, $k$ and $\lambda$, to be abbreviated as BIBD [$b$, $v$, $r$, $k$, $\lambda$]. For optimal estimation of the covariates' effect, therefore, we have to find $\mathbf{Z}$-matrices satisfying (4.3.3) and (4.3.4) with the same $\mathbf{X}_{1d}$ as in RBD but now the structure of $\mathbf{X}_{2d}$ is somewhat different. In respect of $\mathbf{W}$-matrices, the non-zero elements [$\pm 1$] appear only in the $k$ positions in every row and in $r$ positions in every column of the $b \times v$ matrices, precisely corresponding to the incidence matrix of the BIBD. So the situation is more complex than before in the sense that in the case of an RBD, we were to place $\pm 1$'s in all the $bv$ cells of the $\mathbf{W}$-matrices.

Essentially, then, we have to find $\mathbf{W}$-matrices satisfying the conditions presented below, schematically for $1 \le s \le c$,

$$
\begin{array}{rc|c|l}
 & \text{Treatments} & \text{Row totals} & \\
 & & 0 & \\
 & & 0 & \\
\mathbf{W}^{(s)} = & (\pm n_{ij})_{b\times v} & \vdots & \Big\} \text{ Essential conditions} \\
 & & & \quad 1 \le s \le c \\
 & & 0 & \\
\hline
\text{Column totals} & \underbrace{0 \;\; 0 \;\; \cdots \;\; 0}_{\text{Essential conditions}} & 0 & \text{Superfluous condition (Total)}
\end{array}
\qquad (4.3.14)
$$

and, further, for $1 \leq s \neq s' \leq c$,

$$
\begin{array}{rc|cl}
 & \text{Treatments} & \text{Row totals} & \\
\cline{2-2}
 & & 0 & \\
 & & 0 & \\
\mathbf{W}^{(s)} * \mathbf{W}^{(s')} = & (w_{ij}^{(s)} w_{ij}^{(s')}) & \vdots & \left.\right\} \text{Desirable conditions } 1 \leq s < t \leq c \\
 & & 0 & \\
\cline{2-3}
\text{Column totals} & \underbrace{0 \quad 0 \quad \cdots \quad 0} & 0 & \\
 & \text{Desirable conditions} & \text{Essential condition (Total)} &
\end{array}
\qquad (4.3.15)
$$

In (4.3.14) above, $n_{ij}$ can have only the values 0 or 1. In the following, we present construction procedures for finding the **W**-matrices satisfying (4.3.14) and (4.3.15) for a variety of situations involving BIBDs. Recall that a BIBD is said to be symmetric (to be denoted as SBIBD) if we have $b = v$ so that $r = k$.

**Theorem 4.3.3** *Suppose a SBIBD with parameters $b = v$, $r = k$ and $\lambda$ has been constructed by applying Bose's difference technique (Bose 1939), starting with an initial block of size $k$ and developing the same. Suppose, moreover, that a Hadamard matrix $\mathbf{H}_k$ of order $k$ exists. Then we can construct $c^* = (k-1)$ $\mathbf{W}$-matrices satisfying* (4.3.13) *and* (4.3.14).

**Proof.** As before, we consider the Hadamard matrix $\mathbf{H}_k = (\mathbf{h}_1, \mathbf{h}_2, \ldots, \mathbf{h}_{k-1}, \mathbf{1})$. We now construct $\mathbf{W}^{(s)}$ by using the column vector $\mathbf{h}_s$ of $\mathbf{H}_k$ as follows:

Consider the initial block of the SBIBD and display it in the form of the first row vector in the incidence matrix of the SBIBD. Then replace the non-zero elements of this row successively by the elements of the vector $\mathbf{h}_s$ to generate the first row vector of $\mathbf{W}^{(s)}$. Now develop the first row of $\mathbf{W}^{(s)}$ into the full form of $\mathbf{W}^{(s)}$ in as much the same way as one would have generated the full incidence matrix, developing the initial block.

We carry out the above for each of the vectors $\mathbf{h}_1, \mathbf{h}_2, \ldots, \mathbf{h}_{k-1}$. It is now readily verified that all such $\mathbf{W}$-matrices satisfy the conditions (4.3.13)–(4.3.14). □

**Corollary 4.3.3** *Suppose a SBIBD $[b, v, r, k, \lambda]$ has been made available as per the description in Theorem 4.3.3. Suppose further that $\mathbf{H}_m$ exists for some $m$. Then for the BIBD $[B = mb,\ V = v,\ R = mr,\ K = k,\ \Lambda = m\lambda]$, we can construct $c^* = m(k-1)$ $\mathbf{W}$-matrices satisfying* (4.3.14)-(4.3.15).

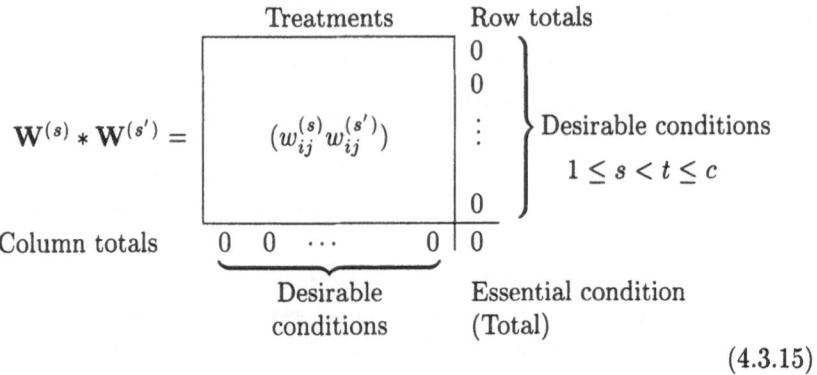

**Proof.** Denote the $\mathbf{W}^{(s)}$-matrices of Theorem 4.3.3 by $\mathbf{W}^{(s)}_{b\times v}$ and the required $\mathbf{W}$-matrices by $\mathbf{G}_{B\times v}$. Based on $\mathbf{W}^{(s)}_{b\times v}$ and the Hadamard matrix $\mathbf{H}_m$ of order $m$, we will construct $m$ $\mathbf{G}_{B\times v}$-matrices as follows:

$$\mathbf{G}^{(s,r)}_{B\times v} = \begin{pmatrix} h^*_{r1}\ \mathbf{W}^{(s)}_{b\times v} \\ h^*_{r2}\ \mathbf{W}^{(s)}_{b\times v} \\ \vdots \\ h^*_{rm}\ \mathbf{W}^{(s)}_{b\times v} \end{pmatrix} = \mathbf{h}^*_r \otimes \mathbf{W}^{(s)}_{b\times v} \tag{4.3.16}$$

Here $\mathbf{H}_m$ is taken as $\mathbf{H}_m = (\mathbf{h}^*_1, \mathbf{h}^*_2, \ldots, \mathbf{h}^*_{m-1}, \mathbf{1}) = (h^*_{rt})$. It is now a routine task to verify the claim of the Corollary. □

**Remark 4.3.2** If a BIBD $(mv, v, mk, k, \lambda)$ is formed by developing $m$ initial blocks each of size $k$, then $c^* = m(k-1)$ whenever $\mathbf{H}_m$ and $\mathbf{H}_k$ exist. We refer to (4.3.15) for a verification of this statement.

**Special Case:** Yates' orthogonal series OS1$[b = v = s^2+s+1, r = k = s+1, \lambda = 1]$.

Suppose $s+1 = 0 \pmod 4$ so that $\mathbf{H}_{s+1}$ exists. Then there are $c^* = s$ $\mathbf{W}$-matrices satisfying the desired conditions. Again, if $s = 2^p$, $p \geq 1$ so that $s^2 = 0 \pmod 4$, then we can work with the complementary BIBD $[b = v = s^2+s+1,\ r = k = s^2,\ \lambda = s(s-1)]$ and claim that there are $c^* = s^2 - 1$ $\mathbf{W}$-matrices having the desired features.

Next we give two illustrative examples.

**Example 4.3.2** SBIBD $[b = v = 7,\ r = k = 4,\ \lambda = 2]$ is the complementary design of SBIBD $[b = v = 7,\ r = k = 3,\ \lambda = 1]$ which is obtained from Yates' OS1 series by putting $s = 2$. Thus we have 3 $\mathbf{W}$-matrices, each of order $7 \times 7$. Below we exhibit the construction in details.

Note that the initial block is $(0, 3, 5, 6) \bmod 7$ for the desired SBIBD with block size 4. Thus, we first display the incidence matrix

$$\mathbf{N} = \begin{pmatrix} 1 & 0 & 0 & 1 & 0 & 1 & 1 \\ 1 & 1 & 0 & 0 & 1 & 0 & 1 \\ 1 & 1 & 1 & 0 & 0 & 1 & 0 \\ 0 & 1 & 1 & 1 & 0 & 0 & 1 \\ 1 & 0 & 1 & 1 & 1 & 0 & 0 \\ 0 & 1 & 0 & 1 & 1 & 1 & 0 \\ 0 & 0 & 1 & 0 & 1 & 1 & 1 \end{pmatrix}.$$

Next we take

$$\mathbf{H}_4 = \begin{pmatrix} 1 & 1 & 1 & 1 \\ -1 & 1 & -1 & 1 \\ 1 & -1 & -1 & 1 \\ -1 & -1 & 1 & 1 \end{pmatrix}.$$

Therefore, the **W**-matrices are given by

$$\mathbf{W}^{(1)}_{7\times 7} = \begin{pmatrix} 1 & 0 & 0 & -1 & 0 & 1 & -1 \\ -1 & 1 & 0 & 0 & -1 & 0 & 1 \\ 1 & -1 & 1 & 0 & 0 & -1 & 0 \\ 0 & 1 & -1 & 1 & 0 & 0 & -1 \\ -1 & 0 & 1 & -1 & 1 & 0 & 0 \\ 0 & -1 & 0 & 1 & -1 & 1 & 0 \\ 0 & 0 & -1 & 0 & 1 & -1 & 1 \end{pmatrix}$$

$$\mathbf{W}^{(2)}_{7\times 7} = \begin{pmatrix} 1 \quad 0 \quad 0 \quad 1 \quad 0 \quad -1 \quad -1 \\ \text{Develop as ususal} \end{pmatrix}$$

$$\mathbf{W}^{(3)}_{7\times 7} = \begin{pmatrix} 1 \quad 0 \quad 0 \quad -1 \quad 0 \quad -1 \quad 1 \\ \text{Develop as usual} \end{pmatrix}.$$

**Example 4.3.3** SBIBD $[b = v = 13, r = k = 4, \lambda = 1]$. Here $s = 2$ and the initial block is $(0, 1, 3, 9) \pmod{13}$. Here $\mathbf{H}_4$ exists so that we can readily construct 3 **W**-matrices.

**Theorem 4.3.4** *Consider a BIBD $[b, v, r, k, \lambda]$ for which $\mathbf{H}_k$ exists and there are available $t$ **W**-matrices satisfying the desirable conditions. Then for the BIBD $[B = mb, V = mr, K = k, \Lambda = m\lambda]$, we have $c^* = mt$ whenever $\mathbf{H}_m$ exists.*

**Proof.** The verification is based on (4.3.15). □

In the absence of any knowledge about the number of covariates that can be accommodated in a given BIBD, we can always achieve $c^* = k - 1$ whenever the number of blocks is doubled and $\mathbf{H}_k$ exists. This is the content of the following theorem.

**Theorem 4.3.5** *Suppose that an entirely arbitrary BIBD $[b, v, r, k, \lambda]$ exists. Then if $\mathbf{H}_k$ exists, we can construct $c^* = k - 1$ **W**-matrices for the BIBD $[B = 2b, V = v, R = 2r, K = k, \Lambda = 2\lambda]$.*

**Proof.** Let $\mathbf{N}_{b\times v}$ denote the incidence matrix of the former BIBD. Let $\mathbf{H}_k = (\mathbf{h}_1, \mathbf{h}_2, \ldots, \mathbf{h}_{k-1}, \mathbf{1})$. In order to construct $\mathbf{W}^{(s)}_{B\times v}$-matrix, we fill up the non-empty positions in $\mathbf{N}_{b\times v}$ by placing the elements of $\mathbf{h}_s$ successively in each row and in the order the positions appear. This results in a matrix of order $b \times v$. Call it $\mathbf{W}^{(s)}_{b\times v}$. Then

$$\mathbf{W}^{(s)}_{B\times v} = \begin{pmatrix} \mathbf{W}^{(s)}_{b\times v} \\ -\mathbf{W}^{(s)}_{b\times v} \end{pmatrix}.$$

It is now easy to assert the claim. □

## 4.4 Concluding Remarks

In a related paper Weirich (1985) has also considered the problem of estimation of the parameters in a CRD model with a single covariate. In applied design problems, often bounds are given on the number of replications for each treatment level, reflecting the limited availability of some levels. He assumes $b_i \leq n_i \leq B_i$, $i = 1, 2, \ldots, v$ along with the stipulation that at least for some $i, N/v$ is not covered by the allocation range available. Thus the usual symmetric allocation is *not* feasible. It is also assumed that the covariates assume values in $[-1, 1]$. Under these constraints, Weirich (1985) has given necessary and sufficient conditions for the D-optimum designs for inference on (i) the treatment effects only, (ii) the regression parameters only, and (iii) all the parameters. Some finite algorithms generating optimum allocation designs are also derived. It is clear that there is scope for further generalization of the problem in two directions:

(a) extending the ANOVA part to RBD and BIBD set-up;

(b) incorparating more covariates in the model.

As a final point, we want to address the issue of possible use of orthogonal arrays (OA) in the constructional problems we have discussed in this Chapter. Take, for example, the case of a CRD set-up. It is true that the **W**-matrices can be displayed in the form of extended row vectors so that eventually we have an OA of order $c \times N$ formed of the elements $+1$ and $-1$ in each row. Clearly, the condition (4.2.4) is satisfied whenever the OA is of strength 2. However, the point to be noted is the implication of the condition (4.2.3). This means that the OA partitions itself into $v$ sub-arrays of order $c \times R$ and all row totals in each of these sub-arrays are 0.

Our non-trivial constructions indicate that the component sub-arrays do *not* necessarily enjoy the property of OA separately and this is not even required. Thus simple component-wise constructions may be ruled out. For an RBD set-up or for a BIBD set-up, the study of the OA becomes quite complicated (because of block and treatment structures imposed on the OA along with empty cells in case of a BIBD) and it is far from being a routine OA any more. We have *not* studied this aspect of the possible use of OA's. It might lead to a better understanding of the constructional problems and also to some more solutions with enhanced value of $c^*$ in some of the situations. Useful references to Hadamard matrices and orhogonal arrays are Raghavarao (1971), Hedayat and Wallis (1978) and Hedayat *et al.* (1999). Fang *et al.* (2000) have investigated the relationship between uniformity (see Fang and Wang 1994) and orthogonality and presented algorithms for construction of orthogonal and nearly orthogonal arrays (see Ma *et al.* 2000).

## References

Bose, R. C. (1939). On the construction of balanced incomplete block designs. *Annals of Eugenics* **9**, 353–399.

Das, K., Mandal, N. K. and Sinha, Bikas K. (2000). Optimal experimental designs for models with covariates. Submitted to the *Journal of the Statistical Planning and Inference.*

Fang, K.-T., Lin, D. K. J., Winker, P. and Zhang, Y. (2000). Uniform design: Theory and application. *Technometrics* Vol. 42, **3**, 237–248.

Fang, K.-T. and Wang, Y. (1994). *Number-theoretic methods in statistics.* Chapman & Hall, London.

Haggstrom, G. W. (1975). The pitfalls of manpower experimentation. RAND Corporation, Santa Monica, California.

Harville, D. A. (1975). Computing optimum designs for covariate models. In *A Survey of Statistical Designs and Linear Models.* J. N. Srivastava Ed. Amsterdam, North Holland, 209–228.

Hedayat, A. S. and Wallis, W. D. (1978). Hadamard matrices and their applications. *Annals of Statistics* **6**, 1184–1238.

Hedayat, A. S., Sloane, N. J. A. and Stufken, J. (1999). *Orthogonal arrays.* Springer, New York.

Lopes Troya, J. (1982a). Optimal designs for covariates models. *Journal of the Statistical Planning and Inference* **6**, 373–419.

Lopes Troya, J. (1982b). Cyclic designs for a covariate model: *Journal of the Statistical Planning and Inference* **7**, 49–75.

Ma, C.-X., Fang, K.-T. and Liski, E. P. (2000). A new approach in constructing orthogonal and nearly orthogonal arrays. *Metrika* **50**, 255–268.

Raghavarao, D. (1971). *Constructions and combinatorial problems in design of experiments.* Wiley, New York.

Rao, C. R. (1973). *Linear statistical inference and its applications. 2nd* ed., Wiley, New York.

Shah, K. R. and Sinha, Bikas K. (1989). *Theory of optimal designs.* Springer, New York.

Weirich, W. (1985). Optimum designs under experimental constraints for a covariate model and an intra-class regression model. *Journal of the Statistical Planning and Inference* **12**, 27–40.

Wu, C. F. J. (1981). Iterative construction of nearly balanced arrangements I: Categorical covariates. *Technometrics* **23**, 37–44.

# 5

# Stochastic Distance Optimality

## Summary

**Features**

**Model(s):** Discrete design models and regression models
**Experimental domains:** Binary design set-up and $\mathcal{T} = [0, 1]$ and $[-1, 1]$ for continuous regressors
**Optimality criteria:** Maximization of distance optimality functional
**Major tools:** Probability inequalities, Schur convexity, invariance
**Optimality results:** Optimal regression designs, optimal designs under CRD and BIBD set-up
**Thrust:** Non-standard optimality functional, normality of error distribution

This Chapter addresses optimality issues for a non-standard optimality criterion viz., the distance optimality criterion-originally introduced in Sinha (1970). Both discrete and regression design models are studied and specific optimality results are presented. This criterion has gained momentum only recently.

## 5.1 Introduction

Sinha (1970) introduced an optimality criterion in certain treatment design settings. This criterion is in terms of the distance between the parameter and its best linear unbiased estimator (BLUE) in a stochastic sense. Hence it is termed *distance optimality* criterion, abbreviated as *DS-optimality* criterion. Shah and Sinha (1989) briefly introduced this concept in a linear model set-up. Only recently, Liski *et al.* (1998) studied it in continuous design settings. Further, Liski *et al.* (1999) studied its general properties and its connection to the traditional optimality criteria. More recently, Mandal *et al.* (2000) investigated the nature of DS-optimal designs for comparison of a set of test treatments with a control, and Liski and Zaigraev (2001) studied certain generalizations of DS-optimality.

This Chapter is based on the contents of the papers cited above. We give below the preliminaries for the study of DS-optimality. In subsequent sections we present its properties and then study the nature of DS-optimal designs in various settings.

We consider distance optimality under the classical linear model

$$\mathbf{Y} \sim \mathrm{N}(\mathbf{X}\boldsymbol{\beta}, \sigma^2 \mathbf{I}_N), \tag{5.1.1}$$

where the $N \times 1$ response vector $\mathbf{Y} = (Y_1, Y_2, \ldots, Y_N)'$ follows a multivariate normal distribution, $\mathbf{X} = (\mathbf{x}_1, \mathbf{x}_2, \ldots, \mathbf{x}_N)'$ is the $N \times m$ model matrix with $\mathrm{rank}(\mathbf{X}) = m$, $\boldsymbol{\beta} = (\beta_1\ \beta_2\ \ldots\ \beta_m)'$ is the $m \times 1$ parameter vector, $E(\mathbf{Y}) = \mathbf{X}\boldsymbol{\beta}$ is the expectation vector of $\mathbf{Y}$ and $\mathbf{V}(\mathbf{Y}) = \sigma^2 \mathbf{I}_N$ is the dispersion matrix of $\mathbf{Y}$, where $\sigma^2 = V(Y_i)$ and $\mathbf{I}_N$ is the $N \times N$ identity matrix. In the sequel we assume for the sake of simplicity that $\sigma^2 = 1$.

Let $\hat{\boldsymbol{\beta}}_d$ be the least squares estimator (LSE) of $\boldsymbol{\beta}$ in (5.1.1) under $d \in \mathcal{D}$, where $\mathcal{D}$ is a given class of competing designs. Now we search for a design $d^* \in \mathcal{D}$ which maximizes

$$P\left(\|\, \hat{\boldsymbol{\beta}}_d - \boldsymbol{\beta} \| \leq \epsilon\right) \text{ for all } \epsilon > 0$$

and for all $d \in \mathcal{D}$, where $\|\cdot\|$ denotes the Euclidean norm. The idea is to minimize the distance between $\hat{\boldsymbol{\beta}}_d$ and $\boldsymbol{\beta}$ in a stochastic sense. Thus $d^*$ is called DS-optimal whenever

$$P\left(\|\, \hat{\boldsymbol{\beta}}_{d^*} - \boldsymbol{\beta} \| \leq \epsilon\right) \geq P\left(\|\, \hat{\boldsymbol{\beta}}_d - \boldsymbol{\beta} \| \leq \epsilon\right) \text{ for all } \epsilon > 0 \tag{5.1.2}$$

and for any competing design $d \in \mathcal{D}$.

Note that the DS-optimality criterion is defined via the *peakedness* of the distributions of $\hat{\boldsymbol{\beta}}_{d^*}$ and $\hat{\boldsymbol{\beta}}_d$. According to a definition proposed by Birnbaum (1948), a random variable $Y_1$ is more peaked about $\mu_1$ than is a random variable $Y_2$ about $\mu_2$ if

$$P\left(|Y_1 - \mu_1| \leq \varepsilon\right) \geq P\left(|Y_2 - \mu_2| \leq \varepsilon\right) \text{ for all } \varepsilon \geq 0. \tag{5.1.3}$$

When $\mu_1 = \mu_2 = 0$, we simply say that $Y_1$ is more peaked than $Y_2$. This definition was generalized to the multivariate case by Sherman (1955). For $k$–dimensional random vectors $\mathbf{Y}_1$ and $\mathbf{Y}_2$, $\mathbf{Y}_1$ is said to be more peaked than $\mathbf{Y}_2$ if

$$P\{\mathbf{Y}_1 \in \mathcal{A}\} \geq P\{\mathbf{Y}_2 \in \mathcal{A}\} \tag{5.1.4}$$

holds for all convex and symmetric (about the origin) sets $\mathcal{A} \subset \mathbf{R}^k$.

There is also a certain similarity between the DS-optimality criterion and Pitman nearness. However, Pitman nearness is a stochastic criterion for comparison of estimators while DS-optimality criterion is for comparison of designs.

Let us now consider a simple example, a line fit model through the origin. Suppose that $N$ uncorrelated responses

$$Y_{ij} = \beta x_i + e_{ij} \tag{5.1.5}$$

follow a normal distribution, $i = 1, 2, \ldots, n$ and $j = 1, 2, \ldots, N_i$ with expectations and variances

$$E(Y_{ij}) = x_i\beta \quad \text{and} \quad V(Y_{ij}) = \sigma^2, \tag{5.1.6}$$

respectively. Then $(\widehat{\beta} - \beta)^2/V[\widehat{\beta}] = \chi_1^2$ follows the central $\chi^2$-distribution with 1 degree of freedom and

$$P(|\widehat{\beta} - \beta| \le \varepsilon) = P((\widehat{\beta} - \beta)^2 \le \varepsilon^2) = P\Big(\chi_1^2 \le \frac{\varepsilon^2}{\sigma^2}\sum_{i=1}^{n} N_i x_i^2\Big). \tag{5.1.7}$$

Let $\mathcal{T} = [0, 1]$ be the regression range. Then the unique maximum of the probability (5.1.7) is $P\big(\chi_1^2 \le \frac{\varepsilon^2 N}{\sigma^2}\big)$. Thus the optimal design is essentially a one-point design which puts all weight at the point 1. We note in passing that under the model (5.1.5) involving a single parameter, the statements

(i) $P(|\widehat{\beta}_d^* - \beta| \le \varepsilon) \ge P(|\widehat{\beta}_d - \beta| \le \varepsilon)$ for all $\varepsilon > 0$,

(ii) $P(|\widehat{\beta}_d^* - \beta| \le \varepsilon) \ge P(|\widehat{\beta}_d - \beta| \le \varepsilon)$ for some $\varepsilon > 0$,

(iii) $V[\widehat{\beta}_d^*] \le V[\widehat{\beta}_d]$

are equivalent (cf. Stępniak 1989).

We now consider a more general situation where $\boldsymbol{\beta}$ may not be estimable. Let $\mathbf{A}$ be a $k \times m$-matrix and the elements of $\boldsymbol{\theta} = \mathbf{A}\boldsymbol{\beta}$ be estimable linear parametic functions of interest to us. We straight away confine to the class of designs $d$ with such a model matrix $\mathbf{X}$ that provides the BLUE of $\boldsymbol{\theta}$ under the model (5.1.1). In the spirit of (5.1.2) we define now a *DS($\varepsilon$)-optimality criterion* as follows. Let $\widehat{\boldsymbol{\theta}}_{d_1}$ and $\widehat{\boldsymbol{\theta}}_{d_2}$ be the LSE's of $\boldsymbol{\theta}$ in (5.1.1) under the designs $d_1$ and $d_2$, respectively. If for a given $\varepsilon > 0$

$$P(\|\widehat{\boldsymbol{\theta}}_{d_1} - \boldsymbol{\theta}\| \le \varepsilon) \ge P(\|\widehat{\boldsymbol{\theta}}_{d_2} - \boldsymbol{\theta}\| \le \varepsilon), \tag{5.1.8}$$

then the design $d_1$ is at least as good as $d_2$ with respect to the DS($\varepsilon$)-criterion.

A design $d^*$ is said to be *DS($\varepsilon$)-optimal* for the LSE of $\boldsymbol{\theta}$ in the model (5.1.1) if it maximizes the probability $P(\|\widehat{\boldsymbol{\theta}} - \boldsymbol{\theta}\| \le \varepsilon)$. When $d^*$ is DS($\varepsilon$)-optimal for all $\varepsilon > 0$, we say that $d^*$ is *DS-optimal*. In the particular case $k = 1$ the DS-criterion coincides with the DS($\varepsilon$)-criterion for any given $\varepsilon > 0$. Note that according to the usual definition of stochastic ordering for random variables (see Marshall and Olkin 1979, p. 481), $\|\widehat{\boldsymbol{\theta}}_{d_1} - \boldsymbol{\theta}\|$ is stochastically less than $\|\widehat{\boldsymbol{\theta}}_{d_2} - \boldsymbol{\theta}\|$, if the inequality (5.1.8) holds for all $\varepsilon > 0$.

Let $\mathcal{I}(d)$ denote the information matrix of $\boldsymbol{\theta}$ per observation under the given model. Note that we assume $\boldsymbol{\theta}$ to be nonsingularly estimable i.e. rank $(\mathbf{A}) = k$. This ensures that $\mathcal{I}(d)$ is *pd*. Let $\mathbf{P}\mathbf{\Lambda}\mathbf{P}'$ be the spectral decomposition of $\mathcal{I}(d)$, where $\mathbf{P}$ is an orthogonal $k \times k$-matrix and $\mathbf{\Lambda} = \text{Diag}(\lambda_1, \lambda_2, \dots, \lambda_k)$ is the diagonal matrix of the eigenvalues of $\mathcal{I}(\text{d})$ arranged in decreasing order. Define $\mathbf{Z} = \sqrt{N}\mathbf{\Lambda}^{\frac{1}{2}}\mathbf{P}'(\widehat{\boldsymbol{\theta}} - \boldsymbol{\theta})$ so that $\mathbf{Z} \sim \text{N}_k(\mathbf{0}, \mathbf{I}_k)$ and $\mathbf{V}(\widehat{\boldsymbol{\theta}}) = \frac{1}{N}\mathbf{P}\mathbf{\Lambda}^{-1}\mathbf{P}'$.

As

$$P(\|\widehat{\boldsymbol{\theta}} - \boldsymbol{\theta}\|^2 \leq \varepsilon^2) = P\Big(\frac{1}{N}\mathbf{Z}'\mathbf{\Lambda}^{-1}\mathbf{Z} \leq \varepsilon^2\Big) = P\Big(\sum_{i=1}^{k} \frac{Z_i^2}{\lambda_i} \leq \delta^2\Big)$$

for all $\delta = \sqrt{N}\varepsilon > 0$, the DS($\varepsilon$)-criterion $P(\|\widehat{\boldsymbol{\theta}} - \boldsymbol{\theta}\| \leq \varepsilon)$ depends on $\mathcal{I}$ only through its eigenvalues $\boldsymbol{\lambda} = (\lambda_1, \lambda_2, \dots, \lambda_k)'$. We define the criterion function $\psi_\varepsilon$, or equivalently $\psi_\delta$, as

$$\psi_\varepsilon(\mathcal{I}) = P(\|\widehat{\boldsymbol{\theta}} - \boldsymbol{\theta}\|^2 \leq \varepsilon^2) \quad \text{and} \quad \psi_\delta(\boldsymbol{\lambda}) = P\Big(\sum_{i=1}^{k} \frac{Z_i^2}{\lambda_i} \leq \delta^2\Big). \qquad (5.1.9)$$

It is clear that $\psi_\varepsilon(\mathcal{I}) = \psi_\delta(\boldsymbol{\lambda})$ for $\delta = \sqrt{N}\varepsilon > 0$.

As a function of $\delta^2$ the DS($\varepsilon$)-optimality criterion $\psi_\delta(\mathbf{\Lambda})$ is the cumulative distribution function of $\sum_{i=1}^{k} Z_i^2/\lambda_i$ for every fixed $\boldsymbol{\lambda} \in \mathbf{R}_+^k$. A design $d^*$ is DS($\varepsilon$)-optimal for the LSE of $\boldsymbol{\theta}$ in (5.1.1), if for a given $\varepsilon > 0$, $\psi_\varepsilon[\mathcal{I}_d^*] \geq \psi_\varepsilon[\mathcal{I}(d)]$ for all $d$. A design $d^*$ is DS-optimal if it is DS($\varepsilon$)-optimal for all $\varepsilon > 0$.

## 5.2 Properties of the DS-Optimality Criterion

The DS($\varepsilon$)-optimality criterion $\psi_\varepsilon(\mathcal{I})$ is a function from the set of $k \times k$ positive definite matrices into the interval $[0, 1]$. Equally well we may consider the corresponding function from positive eigenvalues of $\mathcal{I}$ into $[0, 1]$:

$$\psi_\delta(\boldsymbol{\lambda}) : \mathbf{R}_+^k \to [0, 1].$$

It follows directly from (5.1.9) that

$$\psi_\varepsilon(a\mathcal{I}) = \psi_{\sqrt{a}\varepsilon}(\mathcal{I}) \quad \text{and} \quad \psi_\delta(a\boldsymbol{\lambda}) = \psi_{\sqrt{a}\delta}(\boldsymbol{\lambda})$$

for all $a > 0$. An essential aspect of the optimality criterion $\psi_\varepsilon$ for given $\varepsilon > 0$ is that it induces an ordering among designs and among the corresponding information matrices of designs. We say that a design $d_1$ is at least as good as $d_2$, relative to the criterion $\psi_\varepsilon$, if $\psi_\varepsilon[\mathcal{I}(d_1)] \geq \psi_\varepsilon[\mathcal{I}(d_2)]$. In this case we can also say that the corresponding information matrix $\mathcal{I}(d_1)$ is at least as good as $\mathcal{I}(d_2)$ with respect to $\psi_\varepsilon$.

### 5.2.1 Isotonicity and Admissibility

The function $\psi_\varepsilon$ conforms to the Loewner ordering in the sense that it preserves the matrix ordering, i.e. $\psi_\varepsilon$ is *isotonic* for all $\varepsilon > 0$. The DS-criterion is isotonic relative to Loewner ordering when

$$\mathcal{I}(d_1) \geq \mathcal{I}(d_2) > \mathbf{0} \Rightarrow \psi_\varepsilon[\mathcal{I}(d_1)] \geq \psi_\varepsilon[\mathcal{I}(d_2)] \text{ for all } \varepsilon > 0. \qquad (5.2.1)$$

The proof of this property can be found in Liski *et al.* (1999).

A reasonable weakest requirement for an information matrix $\mathcal{I}$ is that there be no competing information matrix $\tilde{\mathcal{I}}$ which is better than $\mathcal{I}$ in the Loewner ordering sense. We say that an information matrix $\mathcal{I}$ is *admissible* when every competing moment matrix $\tilde{\mathcal{I}}$ with $\tilde{\mathcal{I}} \geq \mathcal{I}$ is actually equal to $\mathcal{I}$ (cf. Pukelsheim 1993, Chapter 10). A design $d$ is rated admissible when its information matrix $\mathcal{I}(d)$ is admissible. The admissible designs form a complete class (Pukelsheim 1993, Lemma 10.3). Thus every inadmissible information matrix may be improved. If $\mathcal{I}$ is inadmissible, then there exists an admissible information matrix $\tilde{\mathcal{I}} \neq \mathcal{I}$ such that $\tilde{\mathcal{I}} \geq \mathcal{I}$. Since $\psi_\varepsilon$ is isotonic relative to Loewner ordering, DS($\varepsilon$)-optimal designs as well as DS-optimal ones can be found in the set of admissible designs.

### 5.2.2 Schur-Concavity of the DS-Optimality Criterion

The notion of majorization proves useful in a study of the function $\psi_\delta(\boldsymbol{\lambda})$. Majorization concerns the diversity of the components of a vector (cf. Marshall and Olkin 1979, p. 7). Let $\boldsymbol{\lambda} = (\lambda_1, \lambda_2, \ldots, \lambda_k)'$ and $\boldsymbol{\gamma} = (\gamma_1, \gamma_2, \ldots, \gamma_k)'$ be $k \times 1$ vectors and $\lambda_{[1]} \geq \lambda_{[2]} \geq \cdots \geq \lambda_{[k]}$, $\gamma_{[1]} \geq \gamma_{[2]} \geq \cdots \geq \gamma_{[k]}$ be their ordered components. Then a vector $\boldsymbol{\lambda}$ is said to majorize $\boldsymbol{\gamma}$, written $\boldsymbol{\lambda} \succ \boldsymbol{\gamma}$, if $\sum_{i=1}^{m} \lambda_{[i]} \geq \sum_{i=1}^{m} \gamma_{[i]}$ holds for all $m = 1, 2, \ldots, k-1$ and $\sum_{i=1}^{k} \lambda_i = \sum_{i=1}^{k} \gamma_i$.

Majorization provides a partial ordering on $\mathbf{R}^k$. The order $\boldsymbol{\lambda} \succ \boldsymbol{\gamma}$ implies that the elements of $\boldsymbol{\lambda}$ are more diverse than the elements of $\boldsymbol{\gamma}$. Then, for example,

$$\boldsymbol{\lambda} \succ \bar{\boldsymbol{\lambda}} = (\bar{\lambda}, \bar{\lambda}, \ldots, \bar{\lambda})' \text{ for all } \boldsymbol{\lambda} \in \mathbf{R}^k,$$

where $\bar{\lambda} = \frac{1}{k} \sum_{i=1}^{k} \lambda_i$. Functions which reverse the ordering of majorization are said to be Schur-concave (cf. Marshall and Olkin 1979, p. 54). In other words, a function $f(\mathbf{x}) : \mathbf{R}^k \to \mathbf{R}$ is said to be a Schur-concave function if for all $\mathbf{x}, \mathbf{y} \in \mathbf{R}^k$ the relation $\mathbf{x} \succ \mathbf{y}$ implies $f(\mathbf{x}) \leq f(\mathbf{y})$. Thus the value of $f(\mathbf{x})$ becomes greater when the components of $\mathbf{x}$ become less diverse. For further details on Schur-concave functions, see Marshall and Olkin (1979).

We say that *DS($\varepsilon$)-criterion is Schur-concave* if the function $\psi_\delta(\boldsymbol{\lambda})$ defined by the equation (5.1.9) is a Schur-concave function of $\boldsymbol{\lambda} = (\lambda_1, \lambda_2, \ldots, \lambda_k)'$.

**Theorem 5.2.1** *The criterion function $\psi_\delta(\boldsymbol{\lambda})$ defined by (5.1.9) has the following properties:*

*(i)* $\psi_\delta(\boldsymbol{\lambda})$ *is a Schur-concave function of* $\boldsymbol{\lambda}$ *for all* $\delta > 0$.

*(ii)* $\psi_\delta(\boldsymbol{\lambda}) \leq \psi_\delta(\bar{\boldsymbol{\lambda}})$ *holds for all* $\boldsymbol{\lambda} \in \mathbf{R}_+^k$ *and all* $\delta > 0$, *where* $\bar{\boldsymbol{\lambda}} = (\bar{\lambda}, \bar{\lambda}, \ldots, \bar{\lambda})'$ *and* $\bar{\lambda} = \frac{1}{k}\sum_{i=1}^{k}\lambda_i$.

*(iii)* $\psi_\delta(\boldsymbol{\lambda})$ *is a Schur-concave function of* $(\log\lambda_1, \log\lambda_2, \ldots \log\lambda_k)'$ *for all* $\delta > 0$.

*(iv)* $\psi_\delta(\boldsymbol{\lambda}) \leq \psi_\delta(\tilde{\boldsymbol{\lambda}})$ *holds for all* $\boldsymbol{\lambda} \in \mathbf{R}_+^k$ *and all* $\delta > 0$, *where* $\tilde{\boldsymbol{\lambda}} = (\tilde{\lambda}, \tilde{\lambda}, \ldots, \tilde{\lambda})'$ *and* $\tilde{\lambda} = \prod_{i=1}^{k}\lambda_i^{\frac{1}{k}}$.

For a proof of the above Theorem see Liski *et al.* (1999). The following is another interesting result on DS-optimality comparison from Liski *et al.* (1999).

If $\boldsymbol{\lambda}$ and $\boldsymbol{\gamma}$ are the column vectors consisting of the eigenvalues of $\boldsymbol{\mathcal{I}}_1$ and $\boldsymbol{\mathcal{I}}_2$, respectively, arranged in decreasing order, and if $\boldsymbol{\mathcal{I}}$ (*not* necessarily equal to $(1-\alpha)\boldsymbol{\mathcal{I}}_1 + \alpha\boldsymbol{\mathcal{I}}_2$) is an information matrix with a vector of eigenvalues $(1-\alpha)\boldsymbol{\lambda} + \alpha\boldsymbol{\gamma}$, then

$$\psi_\varepsilon[(1-\alpha)\boldsymbol{\mathcal{I}}_1 + \alpha\boldsymbol{\mathcal{I}}_2] \geq \psi_\varepsilon(\boldsymbol{\mathcal{I}}). \tag{5.2.2}$$

Theorem 5.2.1 (iii) is a version of Okamoto Lemma (1960). There are other generalizations of this useful result. We reproduce them below from Sinha (1970) and Liski *et al.* (1998). We omit the proofs altogether. To describe the results along similar fashion, we also state Okamoto's Lemma in a form convenient to us. Below $Z_i$'s are iid $N(0,1)$. Note further that in (5.2.3)–(5.2.11) below, $\lambda$'s are placed in the numerator instead of the denominator. In applications, we use the formulation (5.1.9).

**Lemma 5.2.1 (Okamoto 1960)**

$$P\left(\sum_1^k \lambda_i Z_i^2 \leq \epsilon^2\right) \leq P\left(\lambda\chi_k^2 \leq \epsilon^2\right), \tag{5.2.3}$$

*where* $\chi_k^2$ *refers to a central* $\chi^2$*-variate with* $k$ *degrees of freedom and*

$$\lambda = (\Pi\lambda_i)^{1/k}. \tag{5.2.4}$$

**Generalization 1**:

$$P\left(\sum_1^k \lambda_i Z_i^2 \leq \epsilon^2\right) \leq P\left(\lambda'\chi_{k'}^2 + \lambda''\chi_{k''}^2 \leq \epsilon^2\right), \tag{5.2.5}$$

where

$$k = k' + k'', \quad \lambda' = (\Pi_{i=1}^{k'}\lambda_i)^{1/k'} \text{ and } \lambda'' = (\Pi_{i=k'+1}^{k}\lambda_i)^{1/k''}. \tag{5.2.6}$$

**Generalization 2:**

$$P\left(\lambda_1 Z_1^2 + \lambda_2 Z_2^2 \leq \epsilon^2\right) \leq P\left(\mu_1 Z_1^2 + \mu_2 Z_2^2 \leq \epsilon^2\right) \tag{5.2.7}$$

provided that

$$\lambda_1\lambda_2 \geq \mu_1\mu_2 \text{ and } \max\{\lambda_1, \lambda_2\} \geq \max\{\mu_1, \mu_2\}. \tag{5.2.8}$$

**Generalization 3:**

$$P\left(\lambda_1^*\chi_{v_1}^2 + \lambda_2^*\chi_{v_2}^2 \leq \epsilon^2\right) \leq P\left(\lambda^{**}\chi_{v_1}^2 + \lambda_2^{**}\chi_{v_2}^2 \leq \epsilon^2\right) \tag{5.2.9}$$

where

$$(\lambda_1^*)^{v_1}(\lambda_2^*)^{v_2} > (\lambda_1^{**})^{v_1}(\lambda_2^{**})^{v_2} \tag{5.2.10}$$

and

$$\max\{\lambda_1^*, \lambda_2^*\} > \max\{\lambda_1^{**}, \lambda_2^{**}\} > \min\{\lambda_1^{**}, \lambda_2^{**}\} > \min\{\lambda_1^*, \lambda_2^*\}. \tag{5.2.11}$$

## 5.3 Discrete DS-Optimal Designs

We start with the set-up of a CRD model involving $v$ treatments with a total of $n$ observations. Suppose that the treatment replication numbers are $n_1, n_2, \cdots, n_v$. We write the model as

$$y_{ij} = \mu_i + e_{ij}, \; 1 \leq j \leq n_i, \; 1 \leq i \leq v. \tag{5.3.1}$$

In order to characterize a DS-optimal design for the parameter vector $\mu = (\mu_1, \mu_2, \cdots, \mu_v)'$, we refer to (5.1.1), (5.1.2) and Theorem 5.2.1 (ii). Note that $\lambda_i = n_i$ for each $i$. Since $n = \sum n_i$, it turns out that the *symmetrical allocation* viz., $n_1 = n_2 = \cdots = n_v = n/v$ is uniquely DS-optimal *whenever* $n$ is divisible by $v$. The case when $n$ is not divisible by $v$ needs a further argument to reveal that a *most symmetrical allocation* (MSA) viz., one for which any two allocation numbers differ at the most by unity, is again DS-optimal. The original proof in Sinha (1970) is involved and he succeeded in establishing the result only for the case when $v$ is even and $n = v/2$ $(mod\ v)$. The general case follows from Liski *et al.* (1998). We provide all the necessary arguments below to sketch a complete proof.

(i) The probability expression in the canonical form (5.1.9) can be evaluated by taking the variables two at a time, while conditioning on the rest.

(ii) Assume $n_2 - n_1 > 1$. Using $Z_1$ and $Z_2$ while conditioning on the rest, it follows that

$$P\left(\frac{Z_1^2}{n_1} + \frac{Z_2^2}{n_2} \leq \delta^2\right) < P\left(\frac{Z_1^2}{n_1 + 1} + \frac{Z_2^2}{n_2 - 1} \leq \delta^2\right). \tag{5.3.2}$$

In the above, $\delta$ is a generic notation. Generalization 2 provides a proof of the above result. It is now evident that the most symmetrical allocation is DS-optimal.

**Remark 5.3.1** Recently, Saharay and Bhandari (2001) examined the nature of DS-optimal designs in a CRD set-up for inference on *all* elementary treatment contrasts. This corresponds to a singularly estimable full rank problem and hence one has to argue in a different manner. They succeeded in re-establishing optimality of symmetrical or most symmetrical allocations, depending on whether the total number of observations is divisible by the number of treatments or not. More recently, Shah and Sinha (2001) have shown that the MSA is, in fact, universally optimal (UO). This, of course, implies DS-optimality of the MSA.

Next we consider a block design model. Instead of elaborating on the nature of DS-optimal designs for this and other higher-dimensional settings, we will explain the connection between DS-optimal designs and universally optimal designs in the sense of Kiefer (1975). It turns out that in any design setting, if there exists a design $d^*$ for which the **C**-matrix is completely symmetric and its trace is maximum, then the design $d^*$ is DS-optimal for inference on a full set of orthonormal treatment contrasts. Once more, we refer to (5.1.9). Note that for the parametric functions under consideration, the eigenvalues of the information matrix are the positive eigenvalues of the **C**-matrix. In view of complete symmetry and trace maximization, it follows from Theorem 5.2.1 (ii) that the design $d^*$ maximizes (5.1.9) and hence, $d^*$ is DS-optimal. This covers BIBDs, BBDs, Youden square designs and the *regular* generalized Youden designs, in particular (vide Kiefer 1975).

### 5.3.1 Treatment vs. Control Designs

There has been a growing attention to the problem of treatment vs. control comparisons in the set-up of agricultural, industrial and pharmaceutical experiments. Optimality studies in various design settings have also received adequate attention of the researchers. See Majumdar (1996). In this subsection, we intend to specialize to this problem with respect to the DS-optimality criterion. We undertake a study of characterization of DS-optimal designs in the context of a CRD model while the parametric functions are the contrasts involving a set of test treatments and one control. This problem is particularly interesting since here the parametric functions are not orthogonal.

Let $n_0, n_1, \cdots, n_v$ be the allocation numbers subject to $\sum_{i=0}^{v} n_i = n$. We set $p_i = n_i/n$, $i = 0, 1, 2, \ldots, v$ so that $\sum_{i=0}^{v} p_i = 1$. In the sense of the approximate design theory, we shall seek optimal values of the $p_i$'s so as to satisfy (5.1.2) uniformly in $\epsilon > 0$.

We denote by $\eta$ the parameter vector $(\mu_0 - \mu_1, \mu_0 - \mu_2, \ldots, \mu_0 - \mu_v)'$. It is evident that

$$\mathbf{V}(\hat{\eta}) = \sigma^2 \left( \mathbf{n}^{-\delta} + \frac{\mathbf{J}_{vv}}{n_0} \right), \tag{5.3.3}$$

where

$$\mathbf{n}^{-\delta} = \text{Diag}\left(\frac{1}{n_1}, \frac{1}{n_2}, \cdots, \frac{1}{n_v}\right) \quad \text{and} \quad \mathbf{J}_{vv} = \mathbf{1}_v \mathbf{1}_v', \tag{5.3.4}$$

where $\mathbf{1}_v$ is the $v \times 1$ vector of ones. It is easy to see that the eigenvalues of $\mathbf{V}(\hat{\eta})$ satisfy

$$\prod_1^v \lambda_i = |\mathbf{n}^{-\delta} + \mathbf{J}_{vv}/n_0| = \frac{1}{\prod n_i}\left(\frac{n}{n_0}\right) \geq \left(\frac{v}{n - n_0}\right)^v \frac{n}{n_0}. \tag{5.3.5}$$

Further,

$$\begin{aligned} \lambda_{\max} &\geq \frac{\mathbf{1}_v'(\mathbf{n}^{-\delta} + \mathbf{J}_{vv}/n_0)\mathbf{1}_v}{\mathbf{1}'\mathbf{1}} = \frac{\sum_1^v 1/n_i + v^2/n_0}{v} \\ &\geq \frac{v^2/(n - n_0) + v^2/n_0}{v} = \frac{nv}{n_0(n - n_0)}. \end{aligned} \tag{5.3.6}$$

In (5.3.5) and (5.3.6), "=" holds whenever $n_1 = n_2 = \cdots = n_v = (n - n_0)/v$ for every *fixed* $n_0$.

We now refer to the probability inequality (5.2.9). We set

$$v_1 = 1,\ v_2 = v - 1, \quad \lambda_1^* = \lambda_{\max}, \quad \lambda_2^* = (\Pi\lambda_i/\lambda_{\max})^{1/(v-1)}$$

$$\lambda_1^{**} = nv/n_0(n - n_0) \text{ and } \lambda_2^{**} = v/(n - n_0). \tag{5.3.7}$$

There are two cases. Suppose first that $\lambda_2^* \geq \lambda_2^{**}$. Then an application of Generalization 1 gives

$$\begin{aligned} P\left(\sum_1^v \lambda_i Z_i^2 \leq \epsilon^2\right) &\leq P\left(\lambda_{\max}\chi_1^2 + \lambda_2^*\chi_{v-1}^2 \leq \epsilon^2\right) \\ &\leq P\left(\lambda_1^{**}\chi_1^2 + \lambda_2^{**}\chi_{v-1}^2 \leq \epsilon^2\right). \end{aligned} \tag{5.3.8}$$

Here, the last inequality follows by implication of events.

If on the other hand, $\lambda_2^* < \lambda_2^{**}$, conditions (5.2.10) and (5.2.11) are satisfied and hence, the above result follows from application of Generalization 1 followed by that of Generalization 3. Thus, (5.3.8) holds in all cases.

Thus, for every fixed $n_0$, the allocation $(n_0, \frac{n-n_0}{v}, \cdots, \frac{n-n_0}{v})$ is uniformly (in $\epsilon > 0$) better than $(n_0, n_1, \cdots, n_v)$ for all choices of $n_i$'s subject to $\sum_1^v n_i = n - n_0$. Hence, for a fixed $n_0$, the allocation $(n_0, \frac{n-n_0}{v}, \cdots, \frac{n-n_0}{v})$ is essentially complete in the sense that it is uniformly better than any other allocation $(n_0, n_1, \cdots, n_v)$, $\sum_1^v n_i = n - n_0$.

Denoting the right hand side of (5.3.8) by $P(n, v, n_0, \epsilon)$ we have

$$P(n, v, n_0, \epsilon) = P\left(\frac{nv}{n_0(n - n_0)}\chi_1^2 + \frac{v}{n - n_0}\chi_{v-1}^2 \leq \epsilon^2\right). \tag{5.3.9}$$

According to the approximate design theory, we try to choose $p_0(= n_0/n)$ optimally i.e., in such a way that we maximize

$$P(v, p_0, \epsilon) = P\left(\chi_1^2 + p_0\chi_{v-1}^2 \leq p_0(1-p_0)\epsilon^2\right) \qquad (5.3.10)$$

for *every* given $\epsilon > 0$. The above expression (5.3.10) follows from (5.3.9) where $\epsilon$ is used as a "generic" notation .

We carried out extensive computations to determine the optimum value of $p_0$ i.e., the value of $p_0$ which maximizes the coverage probability $P$ for a given value of $\epsilon^2$. The results of these computations are to be found in Mandal *et al.* (2000).

Since the optimal designs are very much $\epsilon$-dependent, it would be desirable to choose a value of $p_0$ which maximizes $P$ averaged with respect to an appropriate weight function for $\epsilon$. We address this problem in the next subsection.

## 5.3.2 Weighted Coverage Probabilities

We shall consider the coverage probabilities discussed in the previous section. Probabilities are *averaged* with respect to $\varepsilon$ using a probability distribution for $\varepsilon^2$. We shall start with the p.d.f. of $\epsilon^2$ as that of $\chi_2^2$ i.e., $\frac{1}{2}e^{-\varepsilon^2/2}$ for $0 < \varepsilon < \infty$. Subsequently, we shall show that our results can be applied for a large class of density functions for $\varepsilon^2$.

The weighted coverage probability is obtained by integrating the right hand side of (5.3.9) with respect to $\varepsilon$ using the p.d.f. for $\varepsilon^2$ given above. Let $\lambda_1 = v/np_0(1-p_0)$ and $\lambda_2 = v/n(1-p_0)$. The weighted coverage probability is given by

$$\overline{P}_{v,\lambda_1,\lambda_2} = \int_0^\infty P(\lambda_1\chi_1^2 + \lambda_2\chi_{v-1}^2 < \varepsilon^2)e^{-\varepsilon^2/2}d(\varepsilon^2/2). \qquad (5.3.11)$$

This may be written as

$$\begin{aligned}\overline{P}_{v,\lambda_1,\lambda_2} &= \int_0^\infty [\int\int_{\lambda_1\chi_1^2+\lambda_2\chi_{v-1}^2<\varepsilon^2} dF(\chi_1^2)dF(\chi_{v-1}^2)]e^{-\varepsilon^2/2}d(\varepsilon^2/2) \\ &= \int_0^\infty dF(\chi_1^2)\int_0^\infty dF(\chi_{v-1}^2)\int_{T/2}^\infty e^{-\varepsilon^2/2}d(\varepsilon^2/2)\end{aligned}$$

where $T = \lambda_1\chi_1^2 + \lambda_2\chi_{v-1}^2$. This is easily seen to be

$$\int_0^\infty\int_0^\infty e^{-T/2}dF(\chi_1^2)dF(\chi_{v-1}^2) = \frac{1}{2^{v/2}\Gamma(\frac{1}{2})\Gamma(\frac{v-1}{2})}$$

$$\int_0^\infty (\chi_1^2)^{-1/2}e^{-(1+\lambda_1)\chi_1^2/2}d\chi_1^2\int_0^\infty (\chi_{v-1}^2)^{(v-3)/2}e^{-(1+\lambda_2)\chi_{v-1}^2/2}d\chi_{v-1}^2.$$

Evaluating the integrals we get

$$\overline{P}_{v,\lambda_1,\lambda_2} = \{(1+\lambda_1)(1+\lambda_2)^{v-1}\}^{-1/2}. \qquad (5.3.12)$$

In the limiting case where $n$ becomes indefinitely large, $\lambda_1 = \lambda_2 = 0$ and in the limit, $\overline{P} = 1$. Henceforth, we will abbreviate $\overline{P}_{v,\lambda_1,\lambda_2} = \overline{P}$.

It is easy to see that

$$(\overline{P})^{-2} = (1 + \lambda_1)(1 + \lambda_2)^{v-1}. \tag{5.3.13}$$

We regard $\overline{P}^{-2}$ as a function of $p_0$ and try to minimize it with respect to $p_0$. For this, we define $\phi(p_0) = \ln\,(\overline{P}^{-2})$ and set $\phi'(p_0) = 0$. This gives

$$\frac{2p_0 - 1}{p_0^2(1 + \lambda_1)} + \frac{v - 1}{1 + \lambda_2} = 0.$$

On simplification, this gives

$$\Delta\{(2p_0 - 1) + p_0(v - 1)\} + (1 - p_0)\{(2p_0 - 1) + p_0^2(v - 1)\} = 0,$$

where $\Delta = v/n$.

Take limit when $n \to \infty$ i.e. when $\Delta \to 0$. This gives

$$(1 - p_0)\{(2p_0 - 1) + p_0^2(v - 1)\} = 0. \tag{5.3.14}$$

It can be easily seen that the only admissible root of the equation (5.3.14) is $p_0 = (1 + \sqrt{v})^{-1}$ .

We now consider the case when $\Delta > 0$. Note that ignoring higher powers of $\Delta$, we have the approximate relation:

$$(\overline{P})^{-2} = (\Delta + 1 - p_0)(\Delta(v - 1) + p_0 - p_0^2)/p_0(1 - p_0)^2.$$

We set a value for $\overline{P}$ and search for a value of $p_0$ which will maximize $\Delta = v/n$ and thereby minimize the value of $n$ required to attain $\overline{P}$.

The table given below gives the optimum values of $p_0$ and the corresponding values of $n$ when $\overline{P} = .9$ or $.95$. We give these for $v = 2, 3, 5, 10$ and 15. We also give the common value of $n_i$, the replication number for the control treatments.

$\overline{P} = .9$

| $v$ | $p_0$ | $n$ | $n_0$ | $n_1$ |
|---|---|---|---|---|
| 2 | 0.409 | 52 | 22 | 15 |
| 3 | 0.360 | 102 | 36 | 22 |
| 5 | 0.304 | 241 | 71 | 34 |
| 10 | 0.236 | 809 | 189 | 62 |
| 15 | 0.202 | 1671 | 336 | 89 |

$\overline{P} = .95$

| $v$ | $p_0$ | $n$ | $n_0$ | $n_1$ |
|---|---|---|---|---|
| 2 | 0.412 | 110 | 46 | 32 |
| 3 | 0.363 | 214 | 79 | 45 |
| 5 | 0.306 | 503 | 153 | 70 |
| 10 | 0.238 | 1676 | 396 | 128 |
| 15 | 0.204 | 3453 | 708 | 183 |

Relative efficiency of two designs having proportions $p_0$ and $p_0^*$ for the test treatment may be measured by the ratio of the values $n$ and $n^*$ required to attain the same average coverage probability $\overline{P}$. This would in fact depend upon $\overline{P}$. However, if we use the approximation used above we get this measure of relative efficiency as

$$e_{p_0,p_0^*} = \frac{p_0(1-p_0)(v-1+p_0^*)}{p_0^*(1-p_0^*)(v-1+p_0)}. \tag{5.3.15}$$

**Remark 5.3.2** We can easily incorporate a scaling factor in the p.d.f. for $\varepsilon$. If the p.d.f. for $\varepsilon$ is $\varepsilon e^{-\varepsilon^2/2\delta}/\delta$, the value of $n$ is given by $n_\delta = n/\delta$. Thus, small values of $\delta$ will lead to larger values of $n_\delta$.

**Remark 5.3.3** If the p.d.f. for $\varepsilon^2$ is a mixture of scaled $\chi_2^2$ p.d.f.'s given by $\frac{1}{2}\sum_i \frac{w_i}{\delta_i} e^{-\varepsilon^2/2\delta_i}$ the coverage probability is given by

$$\overline{P} = \sum_i w_i \delta_i^{v/2}(\delta_i + \lambda_1)^{-1/2}(\delta_i + \lambda_2)^{-(v-1)/2}.$$

For given $w_i$'s and $\delta_i$'s an analysis similar to the above can be carried out. This will be somewhat more complex.

**Remark 5.3.4** If for some design settings $\lambda_1\chi_1^2 + \lambda_2\chi_{v-1}^2$ is replaced by $\lambda_1\chi_{v_1}^2 + \lambda_2\chi_{v_2}^2$, we can use the same technique for computing $\overline{P}$. We will not discuss the set-up of block designs for the study of test vs. control treatment comparisons. A recent study of this topic can be found in Mandal *et al.* (2000).

## 5.4 DS-Optimal Regression Designs

In Section 5.1 we derived the DS-optimal design for the LSE of $\beta$ in the first degree model without intercept on the asymmetric experimental domain $[0, 1]$. Liski *et al.* (1998) obtained the unique DS-optimal design for the LSE of regression coefficients in the first-degree model with intercept on the symmetric domain $[-1, 1]$. We generalize this result for an $m$-factor first-degree model on a symmetric experimental domain.

In Section 5.4.2 we summarize briefly certain general results on symmetric polynomial designs. For a polynomial of degree $k > 1$ there does not exist, in general, a DS-optimal design. This topic is illustrated by using a quadratic model as an example. The classical D- and E- criteria have an interesting relationship with DS($\varepsilon$)-criterion as $\varepsilon$ approaches to 0 and $\infty$, respectively. We present this relationship in Section 5.4.3 where optimal designs for the LSE of $\boldsymbol{\beta}$ are considered in the context of the general linear model (5.1.1).

## 5.4.1 First-Degree Polynomial Fit Models

We consider an $m$-factor first-degree polynomial fit model

$$Y_{ij} = \beta_0 + \beta_1 x_{i1} + \cdots + \beta_m x_{im} + e_{ij} \tag{5.4.1}$$

with $m$ regression variables and a design $d_n$ with $n$ experimental conditions $\mathbf{x}_i = (x_{i1}, x_{i2}, \ldots, x_{im})'$, $i = 1, 2, \ldots, n$ with relative weights $p_1, p_2, \ldots, p_n$ as in (2.4.2). We assume now that the experimental domain is an $m$-dimensional Euclidean ball of radius $\sqrt{m}$, that is $\mathcal{T}_{\sqrt{m}} = \{\mathbf{x} \in \mathbf{R}^m : ||\mathbf{x}|| \leq \sqrt{m}\}$. As in Chapter 2, we use the approximate theory.

Since the DS-optimality criterion is isotonic with respect to the Loewner ordering and it depends on the information matrix only through the eigenvalues, there exists by Theorem 2.4.2 an $(m+1)$-point simplex design in $\mathcal{T}_{\sqrt{m}}$ that dominates any other design with respect to DS-optimality criterion. Therefore the search of DS-optimal design for the LSE of $\boldsymbol{\beta} = (\beta_0, \beta_1, \ldots, \beta_m)'$ can be restricted to the class of $(m+1)$-point simplex designs. As pointed out in Section 2.4.1, there always exists an $(m+1)$-point simplex design for the LSE of $\boldsymbol{\beta}$ in (5.4.1) and any rotation of a simplex design is a simplex design. This observation due to Theorem 2.4.2 greatly simplifies the proof of the following theorem. A corresponding result is given by Liski *et al.* (1999, Theorem 4.2).

**Theorem 5.4.1** *Let $d$ be an $(m+1)$-point design for the LSE of $\boldsymbol{\beta}$ in (5.4.1) over the ball $\mathcal{T}_{\sqrt{m+1}}$. Then $d$ is DS-optimal if and only if it is an $(m+1)$-point uniform simplex design. Such a design always exists and any rotation of it is a uniform simplex design.*

**Proof.** Note that

$$\mathcal{I}(d) = \sum p_i (1, \mathbf{x}_i')'(1, \mathbf{x}_i')$$

as in (2.4.7). As already noted above, a DS-optimal design can always be found in the set of $(m+1)$-point simplex designs. If $d$ is a simplex design, then (vide Section 2.4.1)

$$|\mathcal{I}(d)| \leq (m+1) \prod_{i=1}^{m+1} p_i$$

and by the arithmetic mean -geometric mean inequality

$$\prod_{i=1}^{m+1} p_i \leq \left(\frac{1}{m+1}\right)^{m+1}. \tag{5.4.2}$$

Equality holds in (5.4.2) if and only if $p_1 = p_2 = \cdots = p_{m+1} = \frac{1}{m+1}$ i.e. $d$ is a uniform simplex design. By Theorem 5.2.1 (*iv*)

$$\psi_\varepsilon[\mathcal{I}(d)] \leq \mathrm{P}\left(\chi^2_{m+1} \leq \delta^2(m+1)^{\frac{1}{m+1}}\left(\prod_{i=1}^{m+1} p_i\right)^{\frac{1}{m+1}}\right), \tag{5.4.3}$$

where the right hand side is an increasing function of $\prod_{i=1}^{m+1} p_i$. Thus by (5.4.2) $d$ is a DS-optimal design if and only if it is a uniform simplex design. □

If $m = 1$, then we have the line fit model

$$Y_{ij} = \beta_0 + \beta_1 x_i + e_{ij} \tag{5.4.4}$$

with experimental domain $\mathcal{T} = [-1, 1]$. It follows from Theorem 5.4.1 that the design $d_{\frac{1}{2}} = \{-1, 1; \frac{1}{2}\}$ is the unique DS-optimal design for the LSE of $\boldsymbol{\beta} = (\beta_0, \beta_1)'$ in (5.4.4).

The same result with the help of a different technique was obtained in Liski *et al.* (1998). An $(m+1) \times (m+1)$ matrix $\mathbf{X}$ with entries 1 and $-1$ is called a *Hadamard matrix* if $\mathbf{X}'\mathbf{X} = (m+1)\mathbf{I}_{m+1}$. Thus there exists a two-level uniform simplex design (and therefore a two-level DS-optimal design) with levels $\pm 1$ if and only if there is a Hadamard matrix of order $m+1$. If $m = 2$, for example, there is no Hadamard matrix of order 3 and there exists no two-level uniform simplex design with levels $\pm 1$ in a 2-way first-order polynomial fit model. However, uniform simplex design with exactly 3 support points exists as has been given in Section 2.4 (preceding Theorem 2.4.2). On the other hand, if the experimental domain is a unit square, then the four *extreme* points of the square with equal weights provides *unique* DS-optimal design. Clearly, for such a domain, there does *not* exist any DS-optimal design with 3 support points.

**Corollary 5.4.1** *Let $\mathcal{D}_n$ be the set of designs d with support size $n \geq m+1$ in the m-way first-degree model (5.4.1) on the experimental domain $\mathcal{T}_{\sqrt{m+1}}$. Then a design $d \in \mathcal{D}_n$ is DS-optimal if $\mathcal{I}(d) = \mathbf{I}_{m+1}$.*

The proof is a simple modification of the proof of Theorem 5.4.1 (cf. Liski *et al.* 1999) and it will not be repeated here.

By Theorem 5.4.1, $\mathcal{I}(d) = \mathbf{I}_{m+1}$ holds for a uniform simplex design $d$. Hence a design $d$ satisfying the condition $\mathcal{I}(d) = \mathbf{I}_{m+1}$ exists for the minimal feasible support size $n = m+1$. Existence of a DS-optimal design in general, for any given pair of positive integers $m$ and $n \geq m+1$, seems to be an unsolved problem. Nevertheless, a DS-optimal design can be found for certain values of $m$ and $n \geq m+1$. For example, *the complete factorial design* $2^m$ is a DS-optimal design with $n = 2^m$. The information matrix of the complete factorial design $2^m$ is $\mathbf{I}_{m+1}$, and consequently it is a DS-optimal design for the LSE of $\boldsymbol{\beta}$.

## 5.4.2 Symmetric Polynomial Designs

We posit now the $k$th degree polynomial fit model

$$Y_{ij} = \beta_0 + \beta_1 x_i + \cdots + \beta_k x_i^k + e_{ij},$$

which was introduced in Section 2.3. Let $d$ be a design for the LSE of $\boldsymbol{\beta} = (\beta_0, \beta_1, \ldots, \beta_k)'$ on $\mathcal{T} = [-1, 1]$. The criterion function $\psi_\varepsilon$, given in (5.1.9), is invariant with respect to the reflection operation and the symmetrized design $\bar{d} = \frac{1}{2}(d + d^R)$ is at least as good as $d$ with respect to the criterion $\psi_\varepsilon$ (see Section 2.3.1 and Liski *et al.* 1999). Therefore, in symmetric experimental domains it is sufficient to consider symmetrized designs only.

However, it turns out that in the polynomial regression model of degree $k > 1$ there exists no DS-optimal design for the LSE of $\boldsymbol{\beta}$. We illustrate this by investigating a quadratic regression model

$$\mathrm{E}(Y_{ij}) = \beta_0 + \beta_1 x_i^2 \tag{5.4.5}$$

over the experimental domain $\mathcal{T} = [-1, 1]$. If we denote $z_i = x_i^2$, then the experimental domain of $z$ is $\mathcal{T}_z = [0, 1]$. We may thus seek DS-optimal design for the LSE of $\boldsymbol{\beta} = (\beta_0, \beta_1)'$ over $\mathcal{T}_z = [0, 1]$. Note that $\mathcal{T}_z$ is an asymmetric domain. We will now show that there exists no DS-optimal design for the LSE of $\boldsymbol{\beta}$ on $\mathcal{T}_z = [0, 1]$.

Let $d$ be a design with $n \geq 2$ distinct experimental levels $z_1, z_2, \ldots, z_n$ on $\mathcal{T}_z = [0, 1]$. Then, virtually, Theorem 3.1.1 states that it is always possible to find a two-point design $d_p = \{0, 1; p\}$ such that $d_p$ dominates $d$, where $p$ is weight at the point 0. As DS-optimality criterion is isotonic relative to Loewner ordering, we have the inequality

$$\psi_\varepsilon[\mathcal{I}(d_p)] \geq \psi_\varepsilon[\mathcal{I}(d)] \quad \text{for all} \quad \varepsilon > 0.$$

Therefore $d_p$ is at least as good as $d$ with respect to the DS-optimality criterion. From this it follows that we may restrict ourselves to the class of two-point designs $d_p = \{0, 1; p\}$. The information matrix is of the form

$$\mathcal{I}(d_p) = \begin{pmatrix} 1 & 1-p \\ 1-p & 1-p \end{pmatrix}. \tag{5.4.6}$$

The eigenvalues of $\mathcal{I}(d_p)$ are

$$\lambda_{1,2} = \frac{2-p}{2} \pm \sqrt{\left(\frac{2-p}{2}\right)^2 - p(1-p)}.$$

The greatest eigenvalue $\lambda_1(p)$ is a monotonically decreasing function of $p$, while $\lambda_2(p)$ attains its maximum at $p = \frac{3}{5}$. Let $\mathbf{Z} = \sqrt{N}\Lambda^{\frac{1}{2}}\mathbf{P}'(\widehat{\boldsymbol{\beta}} - \boldsymbol{\beta})$ be defined as in Section 5.1. Since the eigenvalues $\lambda_1(p)$ and $\lambda_2(p)$ of the matrix (5.4.6) are functions of the weight parameter $p$, the DS($\varepsilon$)-optimality criterion $\psi_\varepsilon(\mathcal{I})$ depends on $\mathcal{I}$ solely through $p$. Hence we may write

$$\psi_\delta(\boldsymbol{\lambda}) = \psi_\delta(p) = \mathrm{P}\Big(\frac{Z_1^2}{\lambda_1(p)} + \frac{Z_2^2}{\lambda_2(p)} \leq \delta^2\Big).$$

Note that $\mathbf{Y} \sim \mathrm{N}(\mathbf{X}\boldsymbol{\beta}, \sigma^2\mathbf{I}_N)$, and consequently $\mathbf{Z} = (Z_1, Z_2)' \sim \mathrm{N}_2(\mathbf{0}, \mathbf{I}_2)$. Since $\psi_\delta$ is isotonic relative to Loewner ordering and $\lambda_1(p)$ and $\lambda_2(p)$ are decreasing in $[0, \frac{3}{5}]$, we may conclude that $\psi_\delta(\frac{3}{5}) \geq \psi_\delta(p)$ for all $p \in [0, \frac{3}{5}]$ and all $\delta > 0$. Consequently, if $d_p$ is DS-optimal, then $p \in [\frac{3}{5}, 1)$.

When inquiring DS-optimality of a design $d_p$ we define $U_i = Z_i/(\delta\sqrt{\lambda_i})$. Then $\mathbf{U} = (U_1, U_2)'$ follows a normal distribution $\mathrm{N}_2(\mathbf{0}, \delta^{-2}\boldsymbol{\Lambda}^{-1})$. The criterion $\psi_\delta$ can be written as

$$\begin{aligned}\psi_\delta(p) &= P(\|\mathbf{U}\|^2 \leq 1) \\ &= \frac{\delta^2\sqrt{\lambda_1(p)\lambda_2(p)}}{2\pi} \iint\limits_{\|\mathbf{u}\| \leq 1} \exp\Big\{-\frac{\delta^2}{2}\mathbf{u}'\Lambda\mathbf{u}\Big\}\,\mathbf{du},\end{aligned}$$

where $\|\mathbf{U}\|^2 = U_1^2 + U_2^2$, $\mathbf{u}'\Lambda\mathbf{u} = \lambda_1(p)u_1^2 + \lambda_2(p)u_2^2$ and $\mathbf{du} = du_1 du_2$. The function $\psi_\delta(p)$ is positive and continuous on $(0, 1)$ for all $\delta > 0$ and $\psi_\delta(0) = \psi_\delta(1) = 0$. Hence $\psi_\delta(p)$ has a maximum on the interval $(0, 1)$ at $p^*$, say. However, the maximum point $p^*$ depends on $\delta$. Consequently, there exists no DS-optimal design for the LSE of $\beta$ over $\mathcal{T}_z = [0, 1]$ in the model (5.4.5).

### 5.4.3 Characterization of D- and E-Optimality Using DS($\varepsilon$)-Criterion

Liski *et al.* (1999) studied the behaviour of the DS($\varepsilon$)-criterion, when $\varepsilon$ approaches 0 and $\infty$, respectively. These limiting cases have an interesting relationship with the traditional D- and E-optimality criteria. The results hold in a very general linear model set-up as well.

**Theorem 5.4.2** *Let $\boldsymbol{\lambda}$, $\boldsymbol{\gamma} \in \mathbf{R}^k$ denote vectors whose components are the eigenvalues of the information matrices $\mathcal{I}(d_1)$ and $\mathcal{I}(d_2)$, respectively, arranged in decreasing order. Then the following statements hold:*

(a) *If $\psi_\delta(\boldsymbol{\lambda}) \geq \psi_\delta(\boldsymbol{\gamma})$ for all sufficiently small $\delta > 0$, then $|\mathcal{I}(d_1)| \geq |\mathcal{I}(d_2)|$;*

*if $|\mathcal{I}(d_1)| > |\mathcal{I}(d_2)|$, then $\psi_\delta(\boldsymbol{\lambda}) > \psi_\delta(\boldsymbol{\gamma})$ for all sufficiently small $\delta > 0$.*

(b) *If $\psi_\delta(\boldsymbol{\lambda}) \geq \psi_\delta(\boldsymbol{\gamma})$ for all sufficiently large $\delta$, then $\lambda_k \geq \gamma_k$;*

*if $\lambda_k > \gamma_k$, then $\psi_\delta(\boldsymbol{\lambda}) > \psi_\delta(\boldsymbol{\gamma})$ for all sufficiently large $\delta$.*

For the proof of this theorem see Liski *et al.* (1999).

Note that nothing can be said about the relationship between $\psi_\delta(\boldsymbol{\lambda})$ and $\psi_\delta(\boldsymbol{\gamma})$ for all sufficiently small $\delta > 0$, if $|\mathcal{I}(d_1)| = |\mathcal{I}(d_2)|$ in part (a) of Theorem 5.4.2. Theorem 5.4.2 also shows that the DS($\varepsilon$)-criterion is equivalent to the D-criterion as $\varepsilon \to 0$, and the DS($\varepsilon$)-criterion is equivalent to the E-criterion as $\varepsilon \to \infty$. Thus, loosely speaking, in small balls DS($\varepsilon$)-optimum designs are close to the D-optimum designs and in large balls close to the E-optimum designs. These conclusions are due to the fact that both D- and E-optimal designs are unique (cf. Hoel 1958, Pukelsheim and Studden 1993).

## 5.5 Generalizations of DS-Optimality

The DS-criterion is isotonic relative to Loewner ordering (see Section 5.2.1), that is

$$\mathcal{I}(d_1) \geq \mathcal{I}(d_2) \Rightarrow \psi_\varepsilon[\mathcal{I}(d_1)] \geq \psi_\varepsilon[\mathcal{I}(d_2)] \text{ for all } \varepsilon. \tag{5.5.1}$$

Since $\widehat{\boldsymbol{\theta}} \sim N_k(\boldsymbol{\theta}, \frac{\sigma^2}{n}\mathcal{I}(d)^{-1})$ under the model (5.1.1) and matrix inversion is antitonic in the case of positive definite matrices, i.e.

$$\mathcal{I}(d_1) \geq \mathcal{I}(d_2) \Leftrightarrow \mathcal{I}(d_1)^{-1} \leq \mathcal{I}(d_2)^{-1},$$

the result (5.5.1) is a direct consequence of a well-known corollary from Anderson's theorem on the integral of a symmetric unimodal function over a symmetric convex set (see e.g. Perlman 1989, or Tong 1990, Theorem 4.2.5).

Now we denote $\mathbf{X} \sim N_k(\mathbf{0}, \boldsymbol{\Sigma})$ when a $k \times 1$ random vector $\mathbf{X}$ follows a normal distribution with expectation $\mathrm{E}(\mathbf{X}) = \mathbf{0}$ and dispersion matrix $\mathrm{V}(\mathbf{X}) = \boldsymbol{\Sigma} \geq \mathbf{0}$. Liski and Zaigraev (2001) proved the reverse statement of the above mentioned Corollary, which yields an important characterization of the normal random vectors $\mathbf{X}_1$ and $\mathbf{X}_2$ when their dispersion matrices $\boldsymbol{\Sigma}_1$ and $\boldsymbol{\Sigma}_2$ are in Loewner order $\boldsymbol{\Sigma}_1 \leq \boldsymbol{\Sigma}_2$. We state here without proof this characterization theorem.

**Theorem 5.5.1** *Let* $\mathbf{X}_1 \sim N_k(\mathbf{0}, \boldsymbol{\Sigma}_1)$ *and* $\mathbf{X}_2 \sim N_k(\mathbf{0}, \boldsymbol{\Sigma}_2)$, $k \geq 1$, *where* $\boldsymbol{\Sigma}_1 > \mathbf{0}$. *Then*

$$\mathrm{P}(\mathbf{X}_1 \in A) \geq \mathrm{P}(\mathbf{X}_2 \in A)$$

*for all convex and symmetric (with respect to the origin) sets* $A \subset \mathbf{R}^k$ *if and only if* $\boldsymbol{\Sigma}_1 \leq \boldsymbol{\Sigma}_2$.

We present now a generalization of the DS-optimality criterion due to Liski and Zaigraev (2001). It is called the SC-optimality criterion (S for 'stochastic' and C for 'convex').

Let $\widehat{\boldsymbol{\beta}}_{d_1}$ and $\widehat{\boldsymbol{\beta}}_{d_2}$ be the LSE's of $\boldsymbol{\beta}$ in (5.1.1) under the designs $d_1$ and $d_2$, respectively and let $\mathcal{A}$ be a class of convex symmetric (with respect to the origin) sets in $\mathbf{R}^k$. We say that the design $d_1$ dominates $d_2$

(i) with respect to the $\mathrm{SC}_{\mathcal{A}}$-criterion, if

$$\mathrm{P}\left(\widehat{\boldsymbol{\beta}}_{d_1} - \boldsymbol{\beta} \in A\right) \geq \mathrm{P}\left(\widehat{\boldsymbol{\beta}}_{d_2} - \boldsymbol{\beta} \in A\right) \text{ for all } A \in \mathcal{A}$$

(ii) and with respect to the SC-criterion, if

$$\mathrm{P}\left(\widehat{\boldsymbol{\beta}}_{d_1} - \boldsymbol{\beta} \in A\right) \geq \mathrm{P}\left(\widehat{\boldsymbol{\beta}}_{d_2} - \boldsymbol{\beta} \in A\right)$$

for all convex symmetric (with respect to the origin) sets $A \subset \mathbf{R}^k$.

A design $d^*$ for the LSE of $\boldsymbol{\beta}$ in (5.1.1) is $\mathrm{SC}_{\mathcal{A}}$-*optimal* if

$$P\left(\widehat{\boldsymbol{\beta}}_{d^*} - \boldsymbol{\beta} \in A\right) \geq P\left(\widehat{\boldsymbol{\beta}}_{d} - \boldsymbol{\beta} \in A\right) \text{ for all } A \in \mathcal{A}$$

and for all designs $d$. A design $d^*$ is SC-optimal if it is $\mathrm{SC}_{\mathcal{A}}$-optimal for the class $\mathcal{A}$ of all convex symmetric (with respect to the origin) sets in $\mathbf{R}^k$.

Loewner dominance and SC-dominance induce by Theorem 5.5.1 the same partial ordering among designs. Thus, for example, the set of all admissible designs under Loewner dominance is equivalent to the set of all admissible designs under SC-dominance. Further, a design is Loewner optimal if and only if it is SC-optimal.

By Theorem 3.1.1 the set of designs $d_p = \{0, 1;\ p\}$, $0 < p < 1$, form *a complete class* for the LSE of $\boldsymbol{\beta} = (\beta_0, \beta_1)'$ in (5.4.4) with $[a, b] = [0, 1]$ relative to the Loewner dominance. It follows from Theorem 5.5.1 that the designs $d_p$, $0 < p < 1$, form a complete class also relative to SC-dominance. On the basis of identity (3.1.9) it is also clear that there is no Loewner optimal design, or equivalently, no SC-optimal design for the LSE of $\boldsymbol{\beta}$ in (5.4.4) with $[a, b] = [0, 1]$ as well as with any other $[a, b]$.

Further, the above results yield that any $\mathrm{SC}_{\mathcal{A}}$-optimal is of the form $d_p = \{0, 1;\ p\}$. If we specialize on a certain class $\mathcal{A}$ of convex symmetric (with respect to the origin) sets in $\mathbf{R}^k$, it might be possible to find an optimal design. In fact, DS-optimality is a special case of $\mathrm{SC}_{\mathcal{A}}$-optimality. If $\mathcal{A}$ is taken to be the class of all $k$-dimensional balls centered at the origin, then DS-optimality follows. We know that although there is no SC-optimal design for the LSE of $\boldsymbol{\beta}$ in (5.4.4) with $[a, b] = [-1, 1]$, the design $d_{\frac{1}{2}} = \{-1, 1;\ \frac{1}{2}\}$ is the unique DS-optimal design.

Liski and Zaigraev (2001) present some further results and applications of $\mathrm{SC}_{\mathcal{A}}$-optimality and SC-optimality. However, further research is needed in this new field. Finally, we state without proof a result on multifactor first degree regression models without the intercept term (vide Theorem 2.4.1).

**Theorem 5.5.2** *Let $d$ be an $n$-point design on $\mathcal{T}_{\sqrt{m}}$ for the LSE of $\boldsymbol{\beta} = (\beta_1, \beta_2, \ldots, \beta_m)'$ in* (2.4.1), *$n > m$. Then there exists an $m$-point orthogonal design $d^*$ that dominates $d$ with respect to the SC-optimality criterion.*

However, there is no Loewner optimal or SC-optimal $m$-point orthogonal design for the LSE of $\boldsymbol{\beta}$ in (2.4.5). As shown in Section 5.4.1 a DS-optimal design on $\mathcal{T}_{\sqrt{m}}$ for the LSE of $\boldsymbol{\beta}$ in (2.4.5) always exists. Thus taking as a class of convex sets $\mathcal{A}$ a class of all $m$-dimensional balls centered at the origin yields an optimal design with respect to the $\mathrm{SC}_{\mathcal{A}}$-optimality criterion.

# References

Birnbaum, Z. W. (1948). On random variables with comparable peakedness. *Annals of Mathematical Statistics* **19**, 76–81.

Hoel, P. G. (1958). Efficiency problems in polynomial estimation. *Annals of Mathematical Statistics* **29**, 1134–1145.

Kiefer, J. C. (1974). General equivalence theory for optimum designs (approximate theory). *Annals of Statistics* **2**, 849–879.

Kiefer, J. C. (1975). Construction and optimality of generalized Youden designs. In *A survey of statistical design and linear models.* Ed. J. N. Srivastava. North-Holland, Amsterdam, 333-353.

Liski, E. P., Luoma, A., Mandal, N. K. and Sinha, Bikas K. (1998). Pitman nearness, distance criterion and optimal regression designs. *Calcutta Statistical Association Bulletin* **48** (191-192), 179–194.

Liski, E. P., Luoma, A. and Zaigraev, A. (1999). Distance optimality design criterion in linear models. *Metrika* **49**, 193–211.

Liski, E. P. and Zaigraev, A. (2001). A stochastic characterization of Loewner optimality design criterion in linear models. Appears in *Metrika.*

Majumdar, D. (1996). Optimal and efficient treatment-control designs. In *Handbook of Statistics* **13** (eds. S. Ghosh and C. R. Rao), 1007–1053.

Mandal, N. K. Shah, K. R. and Sinha, Bikas K. (2000). Comparison of test vs. control treatments using distance optimality criterion. *Metrika* **52**, 147-162.

Marshall, A. W. and Olkin, I. (1979). *Inequalities: Theory of majorization and its applications.* Academic Press, New York.

Okamoto, M. (1960). An inequality for the weighted sum of $\chi^2$-variates. *Bulletin of Mathematical Statistics* **9**, 69–70.

Perlman, M. D. (1989), T. W. Anderson's theorem on the integral of a symmetric unimodal function over a symmetric convex set and its applications in probability and statistics. In: *The Collected Papers of T. W. Anderson: 1943–1985.* Vol. 2, Ed. George P. H. Styan. Wiley, New York, 1627–1641.

Pukelsheim, F. (1993). *Optimal design of experiments.* Wiley, New York.

Pukelsheim, F. and Studden, W. J. (1993). E-optimal designs for polynomial regression. *Annals of Statistics* **21**, 402–415.

Saharay, R. and Bhandari, S. (2001). DS-optimal designs in one-way ANOVA. Stat-Math Unit Tech Report no: 3/2001. Indian Statistical Institute, Calcutta. Submitted to *Metrika.*

Shah, K. R. and Sinha, Bikas K. (1989). *Theory of optimal designs.* Springer–Verlag Lecture Notes in Statistics Series, No. 54.

Shah, K. R. and Sinha, Bikas K. (2001). Universal optimality of most symmetric allocation in completely randomized designs. Stat (2001) Conference: Concordia University. July 2001.

Sherman, S. (1955). A theorem on convex sets with applications. *Annals of Mathematical Statistics* **26**, 763–766.

Sinha, Bikas K. (1970). On the optimality of some designs. *Calcutta Statistical Association Bulletin* **20**, 1–20.

Stępniak, C. (1989). Stochastic ordering and Schur-convex functions in comparison of linear experiments. *Metrika* **36**, 291–298.

Tong, Y. L. (1990). *The Multivariate normal distribution.* Springer, New York.

# 6

# Designs in the Presence of Trends

## Summary

**Features**

**Model:** Fixed effects model for block designs with a first degree trend term specific to every block
**Optimality criteria:** Universal optimality (UO)
**Major tools:** Semi balanced arrays, Kiefer's proposition 1
**Optimality results:** Classes of UO designs both within restricted subclasses and within the unrestricted class of designs
**Thrust:** Combinatorial arrangement of treatments within each block of a BIBD so that the trend-adjusted **C**-matrix has maximal trace and completely symmetric structure

In this Chapter we consider a model for a block design where in addition to the block and treatment effect parameters, there is a linear trend term specific to every block. We first give designs which are optimal within the class of binary designs or within the class of trend free designs. We also give designs which are optimal within the unrestricted class. In all cases, we obtain trend-adjusted **C**-matrices which are completely symmetric and have maximal trace so that the designs are universally optimal. Semi balanced arrays of Rao are found very useful in this connection.

## 6.1 Introduction

Often we encounter a trend in experimental material. One could deal with this situation either by adopting appropriate randomization or by choosing a design in which treatment effects (after adjustment for trend) are efficiently estimated. In this Chapter, we shall describe the latter approach.

Bradley and Yeh (1980) first studied the problem of constructing block designs that are "trend-free" in the sense that the information matrix for the treatment effects remains unaltered when adjusted for the trend effect. We shall address the broader problem of obtaining efficient (and, possibly optimal) designs without this restriction. Further, we allow that the trends are possibly different in different blocks.

In Section 6.2, we deal with the preliminaries and have a brief discussion of the estimability of the treatment contrasts in the presence of trend. Section 6.3 deals with universally optimal designs in the restricted class whereas universally optimal designs in the unrestricted class are dealt with in Section 6.4. Much of the material in these two sections is based on the papers by Jacroux *et al.* (1995 and 1997). In all these cases, to get optimal designs, we use Kiefer's well-known result on identification of universally optimal designs in terms of trace maximization and complete symmetry of the resulting information matrices.

## 6.2 Preliminaries

A design $d$ determines the assignment of treatments to the positions within each block. Thus, if $d$ assigns treatment $i$ $(0 \leq i \leq v-1)$ to position $s$ $(1 \leq s \leq k)$ in block $j$ $(1 \leq j \leq b)$ we shall write $d(s,j) = i$. For the observation $y_{sj}$ at position $s$ of block $j$, we consider the model:

$$y_{sj} = \mu + \tau_{d(s,j)} + \beta_j + \phi_1(s)\theta_j + e_{sj} \tag{6.2.1}$$

where $\mu$, $\tau$, $\beta$ and $\theta$ are overall mean, treatment, block and linear trend parameters respectively and $\phi_1$ is the orthogonal polynomial of degree one defined on the positions, assumed to be equi-spaced.

The error term $e_{sj}$ has the properties:

$$E(e_{sj}) = 0,\ V(e_{sj}) = \sigma^2,\ \text{Cov}(e_{sj}, e_{s'j'}) = 0 \ \text{ if } (s,j) \neq (s',j').$$

Bradley and Yeh (1980) assume all $\theta_j$'s to be equal. We waive this restriction outright.

A design $d$ is represented by a $k \times b$ array with entries from $\{0, 1, 2, \ldots, v-1\}$. Here the rows represent the positions whereas the columns represent the blocks. A design $d$ is said to be connected if it permits estimation of all treatment contrasts. The class of such connected designs will be denoted by $\mathcal{D}(v, b, k,)$.

Let $\delta^i_{sj} = \delta^i_{sj}(d) = 1(0)$ if $d(s,j) = i$ $(\neq i)$. We define $n_{d_{ij}} = \sum_s \delta^i_{sj}$, $s_{d_{is}} = \sum_j \delta^i_{sj}$, $r_{d_i} = \sum_j n_{d_{ij}}$ and $h_{d_{ij}} = \sum_s \delta^i_{sj}\ \phi_1(s)$, which is the sum of all $\phi_1(s)$ values corresponding to the treatment $i$ in block $j$. Further,

$$\mathbf{R}_d = \text{Diag}(r_{d_0}, r_{d_1}, \ldots, r_{d_{v-1}}),\quad \mathbf{N}_d = (n_{d_{ij}}),\quad \mathbf{W}_d = (h_{d_{ij}}). \tag{6.2.2}$$

It can be verified that the information matrix for the treatment effects is given by

$$\mathbf{C}_d = \mathbf{R}_d - (1/k)\mathbf{N}_d\mathbf{N}_d' - \mathbf{W}_d\mathbf{W}_d' \tag{6.2.3}$$

If $\theta_j = 0$ for $j = 1, 2, \ldots, b$, we obtain

$$\mathbf{C}_d = \mathbf{C}_{bd} = \mathbf{R}_d - (1/k)\mathbf{N}_d\mathbf{N}'_d.$$

Thus, in general

$$\mathbf{C}_d = \mathbf{C}_{bd} - \mathbf{W}_d\mathbf{W}'_d.$$

We shall denote $\mathbf{W}_d\mathbf{W}'_d$ by $\mathbf{G}_d = (g_{d_{ii'}})$, with $g_{d_{ii'}} = \sum_j h_{d_{ij}} h_{d_{i'j}}$. For the estimability of the treatment contrasts, we note that the $\theta_j$'s are estimable if and only if all treatment contrasts are estimable (assuming that the design is connected when there are no trend effects). This implies that the number of error functions in the design when the trend effects are absent is at least equal to the number of $\theta_j$ parameters i.e., $b$, the number of blocks. Thus, a necessary condition for connectedness in the presence of different trends in different blocks is:

$$bk - b - v + 1 \geq b \text{ or, } bk \geq 2b + v - 1.$$

This also follows from the fact that we are estimating $2b+v-1$ parameters from $bk$ observations.

Further, we note that a linear function of the observations in block $j$ is free from $\beta_j$ and $\theta_j$ if and only if it is orthogonal to the polynomials of the 0th and the first degree i.e., a linear combination of the polynomials of second degree and higher. If a design is a semi balanced array (SBA) i.e., if every unordered pair of treatments occurs equally often in all paired positions $s$ and $s'$ across all blocks (Rao 1973), by the symmetry of the design it follows that the design is connected. Characterization of connected designs beyond this appears to be difficult. It may be noted that SBA's are essentially orthogonal arrays of type II (Rao 1961). We give below an example of an SBA with $v = 5$, $b = 10$, $k = 4$.

| | | | | | | | | | |
|---|---|---|---|---|---|---|---|---|---|
| 1 | 2 | 2 | 3 | 3 | 4 | 4 | 0 | 0 | 1 |
| 2 | 4 | 3 | 0 | 4 | 1 | 0 | 2 | 1 | 3 |
| 3 | 1 | 4 | 2 | 0 | 3 | 1 | 4 | 2 | 0 |
| 4 | 3 | 0 | 4 | 1 | 0 | 2 | 1 | 3 | 2 |

## 6.3 Optimality within Restricted Classes

In this section we shall assume that $k \leq v$.

**(a) Binary Designs:** A design is said to be binary if $n_{d_{ij}} = 0$ or 1, for all $i$, $j$. Let

$$\mathcal{D}_b(v, b, k) = \{d \in \mathcal{D}(v, b, k) |\ n_{d_{ij}} = 0 \text{ or } 1, \text{ for all } i, j\}$$

It is easy to verify that for any bianry design $d$, $\text{tr}(\mathbf{C}_d) = b(k-2)$ i.e. designs in $\mathcal{D}_b(v, b, k)$ have constant trace. We shall now identify some designs in $\mathcal{D}_b(v, b, k)$ which have a completely symmetric (c.s.) information matrix and hence are universally optimal. One way to obtain a completely symmetric $\mathbf{C}_d$ is to choose a balanced incomplete block design (BIBD) and then attempt an allocation which gives c.s. $\mathbf{G}_d$. It is not clear if this is the only way to obtain a c.s. $\mathbf{C}_d$. Since $\text{tr}(\mathbf{G}_d) = b$, $g_{d_{ii}} = b/v$ for a c.s. $\mathbf{G}_d$. This holds for a Youden design, in particular.

Let $m^{ii'}_{ss'}$ denote the number of blocks where the treatments $i$ and $i'$ occur in positions $s$ and $s'$ respectively. Mathematically, for a design $d$,

$$m^{ii'}_{d_{ss'}} = \sum_j \delta^i_{sj}\, \delta^{i'}_{s'j}. \tag{6.3.1}$$

It is easy to see that

$$g_{d_{ii'}} = \sum_s \sum_{s'} \phi_1(s)\, \phi_1(s')[m^{ii'}_{d_{ss'}} + m^{i'i}_{d_{ss'}}] \text{ for i} \neq \text{i}'. \tag{6.3.2}$$

If $m^{ii'}_{d_{ss'}} + m^{ii'}_{d_{s's}} = \alpha$ for all $i \neq i'$, $s < s'$ then (6.3.2) simplifies to $-\alpha$ and hence we get a constant value for $g_{d_{ii'}}$ for all $i \neq i'$. Thus, whenever in a BIBD $m^{ii'}_{d_{ss'}} + m^{ii'}_{d_{s's}}$ is a constant for $i \neq i'$, the design is universally optimal. Recall the definition of SBA which coincides with the constancy of $m^{ii'}_{d_{ss'}} + m^{ii'}_{d_{s's}}$. Since the row sums for $\mathbf{G}_d$ are zero, for an SBA we must have $g_{d_{ii'}} = -b/v(v-1)$ for $i \neq i'$. Since $\phi_1(s)$ is of the form $\phi_1^*(s)/\{k(k^2 - 1)/3\}^{1/2}$ for even values of $k$ (where $\phi_1^*(s)$ is an integer), we must have $bk(k^2-1) = 0 \bmod 3v(v-1)$ wherever $k$ is even. Similarly, we must have $bk(k^2-1) = 0 \bmod 12v(v-1)$ for odd values of $k$. These are the conditions for existence of an SBAfor even and odd values of $k$, respectively.

For $k > 2$, an SBA is in fact a Youden design and hence both $\mathbf{C}_{bd}$ and $\mathbf{G}_d$ are c.s. Further, since $\text{tr}(\mathbf{C}_d)$ is constant for a binary design, it follows that an SBA is universally optimal in $\mathcal{D}_b(v,b,k)$. An example of an SBA with $v = 5$, $b = 10$, $k = 4$ was given in the previous section.

SBA's have been used by several authors such as Morgan and Chakravarti (1988), Cheng (1988), Bhaumik (1993) and Martin and Eccleston (1991) to design experiments with correlated observations. All optimal designs in $\mathcal{D}_b(v,b,k)$ appear to be SBA's or their permutations.

**(b) Trend free designs:** A design $d$ may be called trend free if $\mathbf{C}_s = \mathbf{C}_{bd}$ i.e., $\mathbf{G}_d = 0$, or, equivalently, $\mathbf{W}_d = 0$.

We define

$$\mathcal{D}_{tf}(v,b,k) = \{d \in \mathcal{D}(v,b,k) |\, h_{d_{ij}} = 0 \text{ for } i = 0,1,\ldots,v-1;\ j = 1,2,\ldots,b\}.$$

Since $\phi_1(s) = c\{s - (k+1)/2\}$ for a normalizing constant $c$, $h_{d_{ij}} = 0$ implies $\sum_s s\delta^i_{sj} = n_{d_{ij}}(k+1)/2$. Hence, a binary design cannot be trend free. A necessary condition for a design to be trend free is that

$$n_{d_{ij}}(k+1) = 0 \pmod 2,\ i = 0,1,\ldots,v-1;\ j = 1,2,\ldots,b. \tag{6.3.3}$$

A sufficient condition for (6.3.3) to hold is

$$\delta^i_{sj} = \delta^i_{(k-s+1),j},\ s = 1,2,\ldots,[k/2];\ j = 1,2,\ldots,b;\ i = 0,1,\ldots,v-1. \tag{6.3.4}$$

We note that for $d \in \mathcal{D}_{tf}(v,b,k)$

$$\operatorname{tr} \mathbf{C}_d = bk - (1/k) \sum_j \sum_i n_{d_{ij}}^2. \tag{6.3.5}$$

Hence, in order to maximize $\operatorname{tr} \mathbf{C}_d$ we must minimize $\sum_i \sum_j n_{d_{ij}}^2$ subject to (6.3.3).

We shall first deal with the case where $k$ is even. From (6.3.3) we must have $n_{d_{ij}} \in \{0,2,4,\dots\}$. Since $\sum_i n_{d_{ij}} = k$ to minimize $\sum_i n_{d_{ij}}^2$ we must have $n_{d_{ij}} \in \{0,2\}$ for $i = 0, 1, \dots, v-1$; $j = 1, 2, \dots, b$. From (6.3.4) it is clear that if we start with $d_0 \in \mathcal{D}(v,b,k/2)$ such that the blocks form a BIBD and append to it its *mirror image* $d_0^{-1}$ we obtain a trend free design for which $\mathbf{C}_d$ is c.s. with maximal trace and hence the resulting design is universally optimal. We shall denote the resulting design by $d^* = (d_0, d_0^{-1})'$.

As a final observation, we note that max $\operatorname{tr} \mathbf{C}_d$ occurs when $n_{d_{ij}} = \{0,2\}$ for all $i$, $j$ such that $\sum_i n_{d_{ij}} = k$, $j = 1, 2, \dots, b$. This yields

$$\max_{d \in \mathcal{D}_{tf}(v,b,k)} \operatorname{tr}(\mathbf{C}_d) = b(k-2) = \max_{d \in \mathcal{D}_b(v,b,k)} \operatorname{tr}(\mathbf{C}_d).$$

Thus, *universally optimal binary design* and *universally optimal trend free design* are equally efficient. In fact, the two designs have the same C-matrix since it assumes the form of a multiple of $\mathbf{I} - \mathbf{J}/v$, having the same multiplier.

For odd values of $k$, say $k = 2q - 1$, a simple argument shows that a necessary and sufficient condition for maximizing the trace is:

$$\delta_{qj}^i = 1 \Rightarrow n_{d_{ij}} = 1 \text{ and } \delta_{qj}^i = 0 \Rightarrow n_{d_{ij}} \in \{0,2\}.$$

Based on this, we give the following construction for a universally optimal design in $\mathcal{D}_{tf}(v,b,k)$. We start with $d_1 \in \mathcal{D}(v,b,q-1)$ and $d_0 \in \mathcal{D}(v,b,1)$ such that $d_1$ is a BIBD $(v,b,r,q-1,\lambda)$ and $(d_1,d_0)'$ is a BIBD $(v,b,r,q,\lambda)$.

Then $d^* = (d_1, d_0, d_1^{-1})$ is universally optimal in $\mathcal{D}_{tf}(v,b,k)$ because $\mathbf{C}_{d^*}$ is c.s. with maximal trace. Jacroux *et al.* (1995) have described a method for constructing such designs when $v = 3 \pmod 4$ is a prime number and when $k = v$.

An example of this is the following design in $\mathcal{D}_{tf}(7,7,7)$:

$$d_1 : \left\{ \begin{array}{ccccccc} 3 & 4 & 5 & 6 & 0 & 1 & 2 \\ 5 & 6 & 0 & 1 & 2 & 3 & 4 \\ 6 & 0 & 1 & 2 & 3 & 4 & 5 \end{array} \right.$$

$$d_0 : \quad 0 \;\; 1 \;\; 2 \;\; 3 \;\; 4 \;\; 5 \;\; 6$$

$$d_1^{-1} : \left\{ \begin{array}{ccccccc} 6 & 0 & 1 & 2 & 3 & 4 & 5 \\ 5 & 6 & 0 & 1 & 2 & 3 & 4 \\ 3 & 4 & 5 & 6 & 0 & 1 & 2 \end{array} \right. .$$

Another example is the following design in $\mathcal{D}_{tf}(5,10,5)$.

$$d_1 : \begin{cases} 1 & 1 & 1 & 1 & 2 & 2 & 2 & 3 & 3 & 4 \\ 2 & 3 & 4 & 5 & 3 & 4 & 5 & 4 & 5 & 5 \end{cases}$$

$$d_0 : \quad 5 \;\; 4 \;\; 2 \;\; 3 \;\; 1 \;\; 3 \;\; 4 \;\; 5 \;\; 2 \;\; 1$$

$$d_1^{-1} : \begin{cases} 2 & 3 & 4 & 5 & 3 & 4 & 5 & 4 & 5 & 5 \\ 1 & 1 & 1 & 1 & 2 & 2 & 2 & 3 & 3 & 4 \end{cases}$$

For a design $d^*$ of the above type

$$\begin{aligned} \max_{d \in \mathcal{D}_{tf}(v,b,k)} \operatorname{tr}(\mathbf{C}_d) = \operatorname{tr}(\mathbf{C}_{d^*}) &= b(k-2) + vb/k \\ &> b(k-2) = \max_{d \in \mathcal{D}_b(v,b,k)} \operatorname{tr}(\mathbf{C}_d). \end{aligned}$$

Thus when $k$ is odd, universally optimal trend free design is more efficient than universally optimal binary design. We give here examples for these when $v = 7$, $b = 21$, and $k = 3$.

**Example 6.3.1** For $v > k = 3$, direct computations show that $\operatorname{tr}(\mathbf{C}_d)$ is maximized in $\mathcal{D}(v,b,3)$ if for each $j$, there are two symbols $i$ and $i'$ such that $n_{d_{ij}} = 2$ and $n_{d_{i'j}} = 1$. Hence our design $d^*$ is in fact universally optimal in $\mathcal{D}(v,b,3)$.

However, in general, the optimal design in $\mathcal{D}_{tf}(v,b,k)$ is not optimal in the unrestricted class. To see this we give the following example from Jacroux *et al.* (1995). We have two designs in $\mathcal{D}(7,21,4)$ :

$$d^* = \begin{pmatrix} 0 & 0 & 0 & 4 & 5 & 6 & 1 & 1 & 1 & 5 & 6 & 2 & 2 & 5 & 6 & 3 & 3 & 6 & 4 & 6 & 5 \\ 1 & 2 & 3 & 0 & 0 & 0 & 2 & 3 & 4 & 1 & 1 & 3 & 4 & 2 & 2 & 4 & 5 & 3 & 5 & 4 & 6 \\ 1 & 2 & 3 & 0 & 0 & 0 & 2 & 3 & 4 & 1 & 1 & 3 & 4 & 2 & 2 & 4 & 5 & 3 & 5 & 4 & 6 \\ 0 & 0 & 0 & 4 & 5 & 6 & 1 & 1 & 1 & 5 & 6 & 2 & 2 & 5 & 6 & 3 & 3 & 6 & 4 & 6 & 5 \end{pmatrix}$$

$$\overline{d} = \begin{pmatrix} 0 & 1 & 3 & 1 & 2 & 4 & 2 & 3 & 5 & 4 & 3 & 6 & 5 & 4 & 0 & 6 & 5 & 1 & 0 & 6 & 2 \\ 3 & 0 & 1 & 4 & 1 & 2 & 5 & 2 & 3 & 3 & 6 & 4 & 4 & 0 & 5 & 5 & 1 & 6 & 6 & 2 & 0 \\ 1 & 3 & 0 & 2 & 4 & 1 & 3 & 5 & 2 & 6 & 4 & 3 & 0 & 5 & 4 & 1 & 6 & 5 & 2 & 0 & 6 \\ 0 & 1 & 3 & 1 & 2 & 4 & 2 & 3 & 5 & 4 & 3 & 6 & 5 & 4 & 0 & 6 & 5 & 1 & 0 & 6 & 2 \end{pmatrix}.$$

Both $\mathbf{C}_{d^*}$ and $\mathbf{C}_{\overline{d}}$ are c.s. and $d^*$ is trend free. However, $\operatorname{tr}(\mathbf{C}_{\overline{d}}) > \operatorname{tr}(\mathbf{C}_{d^*})$ and hence $\overline{d}$ is more efficient than $d^*$.

## 6.4 Optimal Designs in $\mathcal{D}(v,b,k)$

In this section we shall attempt to obtain designs which are universally optimal in the entire class $\mathcal{D}(v,b,k)$. We shall first state a number of lemmas which lead to designs with maximal trace. For the proofs of these lemmas, we refer to Jacroux *et al.* (1997).

**Lemma 6.4.1** *Suppose d is a design such that if $n_{d_{ij}} > 0$ in block j, then $n_{d_{ij}} = m$ is a fixed odd number $m \geq 3$. Then*

(a) *If k is odd, the treatments occurring in block j can be arranged in such a way that $h_{d_{ij}} = 0$ for all such treatments.*

(b) *If k is even, it turns out that $|h_{d_{ij}}| \geq \phi_1((k+2)/2)$ for all i such that $n_{d_{ij}} = m$.*
*Further, treatments can be arranged in block j so that $|h_{d_{ij}}| = \phi_1((k+2)/2)$ for all i such that $n_{d_{ij}} = m$.*

Maximizing trace of $\mathbf{C}_d$ requires minimizing trace of $\mathbf{W}_d\mathbf{W}'_d$ [equation (6.2.3)]. The above lemma ensures that this can be accomplished. Equation (6.2.3) also indicates that for trace maximization we need to minimize the differences among the $n_{d_{ij}}$'s in every block. The following lemma shows that this can be accomplished.

**Lemma 6.4.2** *Let $d \in \mathcal{D}(v, b, k)$ be arbitrary. Then we can construct a design $\bar{d}$ having* $\mathrm{tr}(\mathbf{C}_{\bar{d}}) > tr(\mathbf{C}_d)$ *such that*

(a) *If $n_{\bar{d}_{ij}}$ and $n_{\bar{d}_{i'j}}$ are odd, then $n_{\bar{d}_{ij}} = n_{\bar{d}_{i'j}}$;*

(b) *If $n_{\bar{d}_{ij}}$ and $n_{\bar{d}_{i'j}}$ are even, then $|n_{\bar{d}_{ij}} - n_{\bar{d}_{i'j}}| \leq 2$.*

**Lemma 6.4.3** *A design $d \in \mathcal{D}(v, b, k)$ has maximal trace only if for $i = 0, 1, \ldots v-1$ and $j = 1, 2, \ldots, b$*

$$n_{d_{ij}} \in \{\psi, \psi+1, \psi+2\} \text{ for an even integer } \psi.$$

The above lemmas indicate that the desirable values of $n_{d_{ij}}$ are such that all odd values of $n_{d_{ij}}$ are equal whereas the even values differ by at most two.

We now introduce some notations. Let

$$\chi_0(k) = \max\{x \colon \phi_1^2(x) \geq 1/k,\ x \leq k/2\}. \tag{6.4.1}$$

Table 6.1 gives values of $\chi_0(k)$ for some values of $k$

**Table 6.1.** $\chi_0(k)$ for selected values of $k$

| $k$ | 3–6 | 7–11 | 12–16 | 17–21 | 22–26 | 27–31 | 32–36 | 37–40 |
|---|---|---|---|---|---|---|---|---|
| $\chi_0(k)$ | 1 | 2 | 3 | 4 | 5 | 6 | 7 | 8 |

For a non-negative integer $m$, let $f_d(m, j)$ denote the number of treatments that appear $m$ times in block $j$ of design $d$. Thus,

$$f_d(m, j) = |\{i \colon n_{d_{ij}} = m\}|;\ j = 1, 2, \ldots, b. \tag{6.4.2}$$

We describe below allocations of treatments to positions in each block in a design $d^*$ which can be shown to have maximum trace for $\mathbf{C}_d$. For the proof of the result we refer to Jacroux *et al.* (1997).

**Case I:** $3 \le k \le v$. Here each $n_{d^*_{ij}}$ takes values 0, 1 or 2. Further, $f_{d^*}(2,j) = \chi_0(k)$ ($\chi_0$ say). The values of $f_{d^*}(1,j)$ and $f_{d^*}(0,j)$ are obtained from the relations

$$\begin{aligned} f_{d^*}(0,j) + f_{d^*}(1,j) + f_{d^*}(2,j) &= v \\ f_{d^*}(1,j) + 2f_{d^*}(2,j) &= k. \end{aligned} \tag{6.4.3}$$

The $\chi_0$ treatments which appear twice in block $j$ are so arranged that $\delta^i_{sj}(d^*) = \delta^i_{k-s+1,j}(d^*)$, $s = 1, 2, \ldots, \chi_0$ i.e. they are arranged symmetrically at the two ends of the block. The treatments which appear just once are placed in the middle positions in the block.

**Case II:** $v < k < 2v$. Again $n_{d^*_{ij}}$ takes values 0, 1, 2.

$$f_{d^*}(0,j) = \begin{cases} 0, & \text{if } v \le k - \chi_0, \\ v - k + \chi_0, & \text{if } v > k - \chi_0. \end{cases}$$

Values of $f_{d^*}(1,j)$ and $f_{d^*}(2,j)$ are determined from relations (6.4.3). Again, as in case I, treatments which appear twice are placed symmetrically at the two ends of the block whereas the treatments which appear once are placed in the middle positions of the block.

**Case III:** $k \ge 2v$. Here $n_{d^*_{ij}} = [k/v]$ or $[k/v] + 1$ where $[a]$ denotes the largest integer not exceeding $a$. If $k = 0 \pmod v$ and $k/v$ is even, $\delta^i_{sj}(d^*) = \delta^i_{k-s+1,j}$ i.e. the treatments are placed symmetrically. If $k = 0 \pmod v$ and $k/v$ is odd, each block is arranged according to Lemma 6.4.1. If $k \ne 0 \pmod v$, treatments which occur even number of times are placed symmetrically at the two ends whereas the others are placed in the middle positions according to Lemma 6.4.1.

Since for the design $d^*$ with maximum trace the quantities $f_{d^*}(m,j)$ are the same for each $j$, these may be denoted by $f(m)$. The maximized values of $\mathrm{tr}(\mathbf{C}_d)$ for $d \in \mathcal{D}(v,b,k)$ are as follows

(I) $3 \le k < 2v$:

$$\max_{d \in \mathcal{D}(v,b,k)} \mathrm{tr}(\mathbf{C}_d) = b[k - 1 - 2f(2)/k - \sum_{f(2)+1}^{k-f(2)} \phi_1^2(s)].$$

(II) $k$ odd and $\mu v \le k \le (\mu+1)v$ for some integer $\mu \ge 2$:

$$\max_{d \in \mathcal{D}(v,b,k)} \mathrm{tr}(\mathbf{C}_d) = b[k - 2\mu - 1 + \{v\mu(\mu+1)\}/k].$$

(III) $k$ even and $\mu v \le k \le (\mu+1)v$ for some integer $\mu \ge 2$:

$$\max_{d \in \mathcal{D}(v,b,k)} \mathrm{tr}(\mathbf{C}_d) = b[k - 2\mu - 1 + \{v\mu(\mu+1)\}/k - f(\mu_0)\phi_1^2((k+2)/2)],$$

where $\mu_0 \in \{\mu, \mu+1\}$ is odd.

We now attempt to obtain designs for which the $\mathbf{C}_d$ matrix has maximum trace and is c.s. We consider two cases in the following theorems.

**Theorem 6.4.1** *For $3 \le k \le 2v$ let $d$ be the design given by $d = (d_0, d_1^{-1})'$ where $d_0$ is a BIBD $(v, b, r, k^*, \lambda)$ which is an SBA in $b$ columns, $k^*$ rows in $v$ symbols whereas $d_1$ consists of the first $t$ rows of $d_0$ where $t = f(2)$ and $k^* = k - t$. Then the design $d$ is universally optimal in $\mathcal{D}_{(v,b,k)}$.*

**Proof.** For the proof of this and the next theorem, the reader is referred to Jacroux *et al.* (1997).

It may be seen that the design $\bar{d}$ given in Section 6.3 is of the above type and hence is optimal in $\mathcal{D}(7,21,4)$. Another universally optimal design in $\mathcal{D}(7,21,4)$ is given in the above mentioned paper.

For $k \ge 2v$, the following theorem gives some universally optimaldesigns. Let us define $\mu$ as an integer such that $\mu v \le k \le (\mu+1)v$. Let us denote the odd (even) integer in $\{\mu, \mu+1\}$ by $\mu_0(\mu_e)$. We note that

$$f(\mu_0) + f(\mu_e) = v \text{ and } \mu_0 f(\mu_0) + \mu_e f(\mu_e) = k.$$

Thus $f(\mu_0)$ is even when $k$ is even.

**Theorem 6.4.2** *(a) For $k$ odd, let $\eta$ denote BIBD $(v, b, r_0, f(\mu_0), \lambda_0)$ based on $\{0, 1, \ldots, v-1\}$ and suppose that for $j = 1, 2, \ldots, b$, $\eta_j$ denotes the set of treatments in the $j$th block of $\eta$. (If $f(\mu_0) = v$, $\eta$ is a completely randomized block design.) Construct a design $d$ as follows:*

*i) There are $f(\mu_e)$ treatments which are not in $\eta_j$ and each of these occurs $\mu_e$ times in the $j$th block of $d^*$. The positions for these are symmetrically chosen at the two ends of the block i.e. $\delta^i_{sj} = \delta^i_{k-s+1,j}$ for $s = 1, 2, \ldots, \mu_e f(\mu_e)/2$ for $i = 0, 1, 2, \ldots, v-1$.*

*ii) The $f(\mu_0)$ treatments in $\eta_j$ occupy the remaining positions in block $j$ of $d^*$ in accordance with Lemma 6.4.1 (a), each appearing $\mu_0$ times.*

*The design $d^*$ so constructed is universally optimal in $\mathcal{D}(v,b,k)$.*

*(b) For $k$ even, let $\eta$ denote a BIBD $(v, b, r_0, f(\mu_0), \lambda_0)$ based on $\{0, 1, 2, \ldots, v-1\}$ such that $\eta$ is an SBA. (If $f(\mu_0) = 0$, we shall not need $\eta$ in our construction). Let $\eta_j = \eta_j^{(1)} \cup \eta_j^{(2)}$, where $\eta_j^{(1)}$ denotes the set of treatments in rows 1, 2, ..., $f(\mu_0)/2$ of block $j$ of $\eta$ and $\eta_j^{(2)}$ denotes the remaining treatments in block $j$ of $\eta$. We construct $d^*$ as follows:*

*i) The $f(\mu_e)$ treatments not in $\eta_j$ occupy rows 1, 2, ..., $\mu_e f(\mu_e)/2$, $k - \mu_e f(\mu_e)/2 + a, \ldots, k$ each appearing $\mu_e$ times such that $\delta^i_{sj} = \delta^i_{k-s+1,j}$ for $1 = 1, 2, \ldots, \mu_e f(\mu_e)/2$; $i = 0, 1, \ldots, v-1$.*

*ii) The* $f(\mu_0)$ *treatments in* $\eta_j^{(1)} \cup \eta_j^{(2)}$ *occupy the remaining rows of block* $j$ *in accordance with Lemma 6.4.1 (b), each appearing* $\mu_0$ *times such that*

$$h_{d_{ij}} = \begin{cases} -\phi_1(k+2)/2, & \text{if } i \in \eta_j^{(1)}, \\ \phi_1(k+2)/2, & \text{if } i \in \eta_j^{(2)}. \end{cases}$$

*The design* $d^*$ *is universally optimal in* $\mathcal{D}(v,b,k)$.

**Proof.** For the proof of this, the reader is again referred to Jacroux *et al.* (1997).

We give an example of an optimal design in $\mathcal{D}(5,10,14)$. Here $k = 14$ and $v = 5$. We have $\mu_0 = 3$, $\mu_e = 2$, $f(\mu_e) = 1$ and $f(\mu_0) = 4$. For $\eta$, we take

$$\eta = \begin{pmatrix} 1 & 2 & 2 & 3 & 3 & 4 & 4 & 0 & 0 & 1 \\ 2 & 4 & 3 & 0 & 4 & 1 & 0 & 2 & 1 & 3 \\ 3 & 1 & 4 & 2 & 0 & 3 & 1 & 4 & 2 & 0 \\ 4 & 3 & 0 & 4 & 1 & 0 & 2 & 1 & 3 & 2 \end{pmatrix}.$$

The first block of $d^*$ is obtained as follows. The treatment 0 is the only one not in $\eta_1$. This occurs $\mu_e = 2$ times in the first block of $d^*$. We allocate this to the first and the last positions of the block.

Treatments 1 and 2 (3 and 4) are allocated to the remaining twelve positions in such a way that each appears twice and

$$\begin{aligned} h_{d_{ij}} &= -\phi_1(16)/2 \text{ for i} = 1,2, \\ h_{d_{ij}} &= \phi_1(16)/2 \text{ for i} = 3,4. \end{aligned}$$

This gives the first block as follows

| | | | | | | | | | | | | | | |
|---|---|---|---|---|---|---|---|---|---|---|---|---|---|---|
| Position | 1 | 2 | 3 | 4 | 5 | 6 | 7 | 8 | 9 | 10 | 11 | 12 | 13 | 14 |
| Treatment | 0 | 1 | 2 | 3 | 4 | 3 | 4 | 1 | 2 | 4 | 3 | 2 | 1 | 0 |

The remaining blocks of $d^*$ are obtained in a similar manner. Design $d^*$ so obtained is as follows:

$$d^* = \begin{pmatrix} 0 & 0 & 1 & 1 & 2 & 2 & 3 & 3 & 4 & 4 \\ 1 & 2 & 2 & 3 & 3 & 4 & 4 & 0 & 0 & 1 \\ 2 & 4 & 3 & 0 & 4 & 1 & 0 & 2 & 1 & 3 \\ 3 & 1 & 4 & 2 & 0 & 3 & 1 & 4 & 2 & 0 \\ 4 & 3 & 0 & 4 & 1 & 0 & 2 & 1 & 3 & 2 \\ 3 & 1 & 4 & 2 & 0 & 3 & 1 & 4 & 2 & 0 \\ 4 & 3 & 0 & 4 & 1 & 0 & 2 & 1 & 3 & 2 \\ 1 & 2 & 2 & 3 & 3 & 4 & 4 & 0 & 0 & 1 \\ 2 & 4 & 3 & 0 & 4 & 1 & 0 & 2 & 1 & 3 \\ 4 & 3 & 0 & 4 & 1 & 0 & 2 & 1 & 3 & 2 \\ 2 & 4 & 3 & 0 & 4 & 1 & 0 & 2 & 1 & 3 \\ 3 & 1 & 4 & 2 & 0 & 3 & 1 & 4 & 2 & 0 \\ 1 & 2 & 2 & 3 & 3 & 4 & 4 & 0 & 0 & 1 \\ 0 & 0 & 1 & 1 & 2 & 2 & 3 & 3 & 4 & 4 \end{pmatrix}.$$

## 6.5 Efficiency Bounds

When an optimal design cannot be obtained, we may use the following approach to obtain efficiency bounds. We may wish to define the efficiency of a design as

$$e_0(d) = \mathrm{tr}(\mathbf{C}_{d_0}^+)/\,\mathrm{tr}(\mathbf{C}_d^+),$$

where $\mathbf{A}^+$ denotes the Moore–Penrose inverse of $\mathbf{A}$ and $d_0$ has minimal trace $\mathbf{C}_d^+$ over all $d \in \mathcal{D}(v,b,k)$. One difficulty in computing $e_0(d)$ is that $\mathrm{tr}(\mathbf{C}_{d_0}^+)$ may not be known for a given class $\mathcal{D}(v,b,k)$. A lower bound for $e_0(d)$ may be obtained by noting that

$$\mathrm{tr}(\mathbf{C}_{d_0}^+) \geq (v-1)^2/\,\mathrm{tr}(\mathbf{C}_{d_0}) \geq (v-1)^2/\,\mathrm{tr}(\mathbf{C}_{d_1}),$$

where $d_1$ maximizes $\mathrm{tr}(\mathbf{C}_d)$ over $\mathcal{D}(v,b,k)$. The result follows from the relation between the harmonic mean and the geometric mean. Thus, for a given design $d \in \mathcal{D}(v,b,k)$,

$$e_0(d) \geq (v-1)^2/\{\mathrm{tr}(\mathbf{C}_{d_1})\,\mathrm{tr}(\mathbf{C}_d^+)\} = \bar{e}_0(d).$$

Further, if $\mathbf{C}_d$ is c.s., $\mathrm{tr}(\mathbf{C}_d^+) = (v-1)^2/\,\mathrm{tr}(\mathbf{C}_d)$ and hence $\bar{e}_0$ reduces to $\mathrm{tr}(\mathbf{C}_d)/\,\mathrm{tr}(\mathbf{C}_{d_1})$. If $\mathbf{C}_{d_0}$ is c.s., then $\mathrm{tr}(\mathbf{C}_{d_0})$ is maximal and if further $\mathbf{C}_d$ is c.s., then $e_0(d) = \bar{e}_0(d)$. Since $\mathrm{tr}(\mathbf{C}_{d_1})$ can be computed exactly, $\bar{e}_0(d)$ provides a useful lower bound for $e_0(d)$.

**Example 6.5.1** Let $v = 7$, $k = 4$ and $b = 21$. Then $\overline{\max}\ \mathrm{tr}\,\mathbf{C}_d = 50.4$ and it is attained by the design $\bar{d}$ given in 6.3. Clearly, $\bar{d}$ is universally optimal. For the design $d^*$ given in Section 6.3, $\mathrm{tr}\,\mathbf{C}_d = 42$. Hence the efficiency of $d^*$ is $42/50.4 = 83.3\%$.

Majumdar (1998) has given an alternative derivation for the results of this chapter when $k \leq v$. His proof is based on the similarity of the structure of the $\mathbf{C}_d$ matrix for designs with trend and for the design with correlated responses. He then uses the results of Cheng (1988) and Martin and Eccleston (1991) for designs with correlated responses. He also gives optimal arrangements for quadratic and cubic trends.

More recently, Chai and Majumdar (2000) have obtained optimal designs for a model with a random trend effect.

# References

Bhaumik, D. K. (1993). On optimal block designs in the presence of a linear trend. *Sankhyā* B **55**, 91–102.

Bradley, R. A. and Yeh, C. M. (1980). Trend-free block designs: Theory. *The Annals of Statistics* **8**, 883–893.

Chai, F-Shun and Majumdar, D. (2000). Optimal designs for nearest-neighbor analysis. *Journal of Statistical Planning and Inference* **86**, 265-275.

Cheng, C. S. (1988). A note on the optimality of semibalanced arrays, in *Optimal Design and Analysis of Experiments*, eds. Y. Dodge, V. Federov, and H. Wynn. North-Holland, Amsterdam, 115–122.

Jacroux, M., Majumdar, D., and Shah, K. R. (1995). Efficient block designs in the presence of trends, *Statistica Sinica* **5**, 605–615.

Jacroux, M., Majumdar, D., and Shah, K. R. (1997). On the determination and construction of optimal block designs in the presence of linear trends, *Journal of the American Statistical Association* **92**, 375–382.

Majumdar, D. (1998). Design and analysis in the presence of trends. Read at the conference *Experimental Design: Theory and Applications* held at Mathematics Research Institute, Oberwolfach, Germany.

Martin, R. J. and Eccleston, J. A. (1991). Optimal block designs for general dependence structures, *Journal of Statistical Planning and Inference* **28**, 67–81.

Morgan, J. P. and Chakravarti, I. M. (1988). Block designs for first and second order neighbor correlations, *Annals of Statistics* **16**, 1206–1224.

Rao, C. R. (1961). Combinatorial arrangements analogous to OA's, *Sankhya A* **23**, 283–286.

Rao, C. R. (1973). Some combinatorial problems of arrays and applications to experimental designs, in *A survey of Combinatorial Theory*, eds. J. N. Srivastava, F. Harary, C. R. Rao, G. C. Rota and S. S. Srikhande, North-Holland, Amsterdam, 349–359.

# 7

# Additional Selected Topics

## Summary

**Features**

**Models:** Competing effects model, split block model, nested model, 3-way classification model, Poisson count model
**Optimality criteria:** Universal optimality (UO), A- D- and E-optimality
**Major tools:** Complete symmetry and trace maximization [Kiefer's Proposition 1]
**Optimality results:** Characterization of UO designs in different settings
**Thrust:** Diverse applications of UO and balancing

In this Chapter, we dwell on some design settings and present the underlying optimal designs, covering UO or specific optimality criteria viz., A-, D- and E-optimality. The purpose is to acquaint the readers with a variety of interesting and non-standard application areas of optimal designs. We specially mention (i) competing effects model, (ii) split block designs and (iii) models with heteroscedastic errors. In addition, (iv) nested designs and (v) 3-way balanced designs are also discussed.

## 7.1 Introduction

In this Chapter, we shall deal with a number of different settings and discuss optimality results in each of the settings. We deferred a discussion on universal optimality (UO) so long and now we intend to present a slightly modified formulation of it which will be used in some sections to follow. Details of these considerations on UO can be found in Shah and Sinha (2001a).

Let $\mathbf{C}_d$ denote the information matrix for a design $d$ which is used for comparing a set of $v$ treatments in a given design setting. When the context is clear, the suffix $d$ will be dropped. A design minimizing $\Phi(\mathbf{C})$ with respect to every criterion satisfying the following will be called UO:

(i) $\Phi(\mathbf{C}_g)$ is the same for all $g$ where $g$ is a permutation on $\{1, 2, \ldots, v\}$ and $\mathbf{C}_g$ is obtained from $\mathbf{C}$ by applying the permutation $g$ to the rows and columns of $\mathbf{C}$.

(ii) If $\mathbf{C}_1 - \mathbf{C}_2$ is nnd, then $\Phi(\mathbf{C}_1) \leq \Phi(\mathbf{C}_2)$.

(iii) If $\Phi(\mathbf{C}_1) > \Phi(\mathbf{C}_2)$, then $\Phi(t\mathbf{C}_1) > \Phi(t\mathbf{C}_2)$ for t = 2, 3, ....

(iv) $\Phi(\sum t_g \mathbf{C}_g) \leq \Phi((\sum t_g)\mathbf{C})$ where $t_g$'s are non-negative integers.

Sometimes (iv) is replaced by a weaker condition

(iv)′ $\Phi(\sum \mathbf{C}_g) \leq \Phi(v!\mathbf{C})$.

We may call (iv)′ the *symmetry* condition.

It is well-known (Kiefer 1975, Shah and Sinha 1989) that $d^*$ is UO if it satisfies the following conditions:

(a) $\mathbf{C}_{d^*}$ is completely symmetric (c.s.) i.e., $\mathbf{C}_{d^*} = p\mathbf{I} + q\mathbf{J}$ for some scalars $p$ and $q$ satisfying $p + vq = 0$.

(b) $\mathrm{tr}(\mathbf{C}_{d^*}) \geq \mathrm{tr}(\mathbf{C}_d)$ for every competing design $d$.

Yeh (1986) replaced condition (a) by the following condition:

(a)′ For any competing design $d$, there exists a set of positive integers $a_g$ such that $(\sum a_g)\mathbf{C}_{d^*} = \sum a_g \mathbf{C}_g$.

The above is a statement of Yeh's result suitably modified for the present formulation. In fact, we can combine the two results in the form of the following theorem.

**Theorem 7.1.1** *If for any design d, we have a set of non-negative integers $a_g$ such that*

$$\left(\sum a_g\right)\mathbf{C}_{d^*} - \sum a_g \mathbf{C}_g \text{ is nnd,} \tag{7.1.1}$$

*then $d^*$ is UO.*

**Proof.** The inequality

$$\Phi\left(\left(\sum a_g\right)\mathbf{C}_{d^*}\right) \leq \Phi\left(\sum a_g \mathbf{C}_g\right)$$

follows by (ii), and

$$\Phi\left(\sum a_g \mathbf{C}_g\right) \leq \Phi\left(\left(\sum a_g\right)\mathbf{C}_d\right)$$

by (iv). Hence, by (iii),

$$\Phi(\mathbf{C}_{d^*}) \leq \Phi(\mathbf{C}_d).$$

□

The above result does not depend upon the class of competing designs. The term "universal" in UO means universality with respect to optimality criteria and not with respect to the class of competing designs which is regarded as fixed. In fact, the above result can be used to compare the designs $d$ and $d^*$ in the sense that (7.1.1) is a sufficient condition for removing $d$ from the set of designs competing with $d^*$. Primary motivation due to Kiefer (1975) for this approach was to prove optimality with respect to a very broad class of optimality criteria which was intended to include almost all the known criteria. We note that in order to compare two designs under the distance optimality criterion used in Chapter 5, we need to replace (7.1.1) by

$$v!\mathbf{C}_{d^*} - \sum \mathbf{C}_g \text{ is nnd.} \tag{7.1.2}$$

## 7.2 Optimal Designs for a Competing Effects Model

Raghavarao *et al.* (1986) introduced a model where in a block design every observation is affected by "competition" effects from other available treatments in the same block. Such a model arises in a very natural manner in marketing situations where the sale of a brand is affected by the other brands present at the location.

Bhaumik (1995) established UO of 3-designs (to be formally defined below) for the estimation of the competition effects contrasts and for the joint estimation of the treatment effects contrasts and the competition effects contrasts. Bhaumik considered optimality within a class of designs with constant block sizes. However, his application of the permutation group was *not* correct. Later, Raghavarao and Zhou (1998) showed that designs with constant block size do *not* give non-singular information matrices and hence do *not* give full estimability of all the effects. They established UO of unequal block size 3-designs. Our treatment here is largely based on the work of the above authors.

Following Raghavarao and Zhou, we shall consider $D(b, v, k_1, k_2, \ldots, k_b)$, the class of binary block designs to accommodate $v$ treatments in $b$ blocks of sizes $k_1, k_2, \ldots, k_b$ respectively. For a design $d$, let $S_{t(d)}$ denote the set of treatments in the $t$th block. When $d$ is understood, we will simply write $S_t$. For every $i \in S_t$, we postulate the model

$$y_i(S_t) = \tau_i + \sum_{\substack{j \in S_t \\ j \neq i}} \gamma_{i(j)} + e_i(S_t);\ i \in S_t,\ t = 1, 2, \ldots, b \tag{7.2.1}$$

where $y_i$ is the observation recorded on the $i$th treatment in block $t$, $\tau_i$ is the effect of the $i$th treatment, $\gamma_{i(j)}$ is the competition effect of the $j$th treatment on the $i$th treatment and $e_i(S_t)$'s are random errors assumed to be uncorrelated and to have a common variance $\sigma^2$.

It would be tempting to include a term $\beta_t$ for the block effect but it is readily seen that these effects are confounded with the competition effects. A precise argument would be as follows. For every $i$, contrasts involving the effects $[\gamma_{i(j)}|j(\neq i)]$ would be estimated only from observational contrasts involving $i$th treatment across all blocks. Thus, if $i$ belongs to both the sets $S_t$, $S_{t'}$, $t \neq t'$, then $\beta_t - \beta_{t'}$ would be confounded with $(k_t - k_{t'})\bar{\gamma}_i$ and appropriate contrasts involving $\gamma_{i(j)}$'s for $j(\neq i)$ belonging to both $S_t$ and $S_{t'}$ $[\bar{\gamma}_i = \sum_{j\neq i} \gamma_{i(j)}/(v-1)]$. Hence blocking in the true sense would complicate the whole issue. In the above, $\beta$'s represent block effects, and are therefore, taken to be zero in our subsequent analysis.

We now introduce some notations for the design $d$, again dropping the suffix $d$. Let $r_i$ denote the number of blocks containing treatment $i$ and $\lambda_{ij}$ denote the number of blocks containing treatments $i$ and $j$. Further, let $\delta_{ijm}$ denote the number of blocks involving the treatments $i$, $j$, and $m$. We also define

$$\mathbf{r} = (r_1,\ r_2,\ \cdots,\ r_v)'; \quad \mathbf{r}^\delta = \text{Diag}(r_1, r_2, \ldots, r_v);$$
$$\boldsymbol{\lambda}_i = (\lambda_{i1},\ \lambda_{i2},\ \ldots,\ \lambda_{ii-1},\ \lambda_{ii+1},\ \ldots,\ \lambda_{iv})',$$
$$\boldsymbol{\gamma}_{(i)} = (\gamma_{i(1)},\ \gamma_{i(2)},\ \cdots \gamma_{i(i-1)},\ \gamma_{i(i+1)},\ \cdots,\ \gamma_{i(v)})', \quad i = 1, 2, \ldots, v;$$

$$\mathbf{M}_{(i)} = (\delta_{ijm}); \quad i = 1, 2, \ldots, v; \quad i \neq j \neq m. \tag{7.2.2}$$

We stipulate that the $j$th diagonal element of $\mathbf{M}_{(i)}$ is $\lambda_{ij}$. Further, let

$$\boldsymbol{\beta}_{(i)} = (\tau_i,\ \boldsymbol{\gamma}'_{(i)})'; \quad \boldsymbol{\beta} = (\boldsymbol{\beta}'_{(1)},\ \boldsymbol{\beta}'_{(2)},\ \ldots,\ \boldsymbol{\beta}'_{(i)},\ \ldots,\ \boldsymbol{\beta}'_{(v)})'. \tag{7.2.3}$$

It is easy to see that the expectations of the $r_i$ observations on treatment $i$ involve parameters $\boldsymbol{\beta}_{(i)}$ only. We now consider the difference between two observations on treatment $i$ coming from blocks $t$ and $t'$. This has expected value

$$\sum_{j \in S_t} \gamma_{i(j)} - \sum_{j' \in S_{t'}} \gamma_{i(j')} \tag{7.2.4}$$

and we can express this as

$$\sum_{j \in S_t} (\gamma_{i(j)} - \bar{\gamma}_i) - \sum_{j' \in S_{t'}} (\gamma_{i(j')} - \bar{\gamma}_i) + (k_t - k_{t'})\bar{\gamma}_i, \tag{7.2.5}$$

where, as before, $\bar{\gamma}_i = \sum_{j\neq i} \gamma_{i(j)}/(v-1)$.

Thus, we see that if all the blocks involving treatment $i$ are of the same size, we can *not* estimate $\bar{\gamma}_i$. (Note that $\gamma_{i(j)} - \bar{\gamma}_i$ involves only the contrasts in the $v$-1 elements of $\gamma_{(i)}$. Thus a necessary condition for full estimability of $\boldsymbol{\beta}_{(i)}$ is that $i$th treatment occurs in at least two blocks of unequal sizes.)

The information matrix for $\boldsymbol{\beta}_{(i)}$ is seen to be

$$\mathbf{C}_{d(i)} = \mathbf{H}_i = \begin{pmatrix} r_i & \lambda_{i1} & \lambda_{i2} & \dots & \lambda_{iv-1} & \lambda_{iv} \\ \lambda_{i1} & \lambda_{i1} & \delta_{i12} & \dots & \delta_{i1v-1} & \delta_{i1v} \\ \vdots & \vdots & \vdots & \ddots & \vdots & \vdots \\ \lambda_{iv} & \delta_{iv1} & \delta_{iv2} & \dots & \delta_{ivv-1} & \lambda_{iv} \end{pmatrix} \tag{7.2.6}$$

The argument after (7.2.5) suggests that rank $(\mathbf{H}_i) \le v$ for the cases when at least two blocks are of unequal sizes while rank $(\mathbf{H}_i) \le v-1$ in case all the blocks have the same size. This holds independently of the nature of the underlying designs.

Further, the information matrix $\mathbf{C}_d$ for $\boldsymbol{\beta}$ is a block diagonal matrix where the $i$th block diagonal is $\mathbf{H}_i$. We now formally define a 3-design. A design $d^*$ in $D(b, v, k_1, k_2, \dots, k_b)$ is said to be a 3-design if it satisfies the following conditions: (i) $r_i$'s are all equal; (ii) $\lambda_{ij}$'s are all equal; (iii) $\delta_{ijm}$'s are all equal. For a 3-design, we will use the notations: $\mathbf{H}_i^*$, $\mathbf{C}_{d^*(i)}$; $i = 1, 2, \dots, v$; $\mathbf{C}_{d^*}$.

At this stage, we separate the cases of unequal block sizes and of equal block size.

**Case I:** $k_1, k_2, \dots, k_b$ are not all equal

For this case, Raghavarao and Zhou (1998) show that for a 3-design, the information matrices $\mathbf{H}_i^*$ are all identical and positive definite. Thus if a 3-design contains at least two blocks of unequal sizes, one $\mathbf{H}_i^*$ and hence every $\mathbf{H}_i^*$ turns out to be positive definite. Consequently, every treatment occurs in blocks of unequal sizes. We assume that a 3-design $d^*$ exists.

Since there are altogether $v^2$ parameters in the $\boldsymbol{\beta}_{(i)}$'s as a whole, we need at least $v^2$ observations for estimability and, that too, in at least two blocks of unequal sizes. Here is an example of a 3-design with just $v^2$ observations

$$b = 1 + \frac{v(v-1)}{2}, \quad k_1 = k_2 = \dots = k_{b-1} = 2, \quad k_b = v;$$
$$\text{Blocks: } \{(i,j) | 1 \le i < j \le v\} \text{ and } (1, 2, \dots, v).$$

Let $g$ be a permutation on $\{1, 2, \dots, v\}$ and let $\mathbf{G}_g$ be a $v^2 \times v^2$ matrix such that $\mathbf{G}_g'\mathbf{C}\mathbf{G}_g = \mathbf{C}_g$ is the matrix obtained by re-labeling the treatments 1, 2, ..., $v$ according to the permutation $g$. Raghavarao and Zhou (1998) show that for *any* $d \in D(\text{v}, b, k_1, k_2, \dots, k_b)$,

$$v!\,\text{Diag}(\mathbf{H}^*, \mathbf{H}^*, \dots, \mathbf{H}^*) = \sum_g \mathbf{C}_{dg}, \tag{7.2.7}$$

where $\mathbf{C}_{dg} = \mathbf{G}_g'\mathbf{C}_d\mathbf{G}_g$ and $\mathbf{H}^*$ is the common $\mathbf{H}$-matrix for the 3-design. Thus, by Theorem 7.1.1 (appropriately modified for the present set-up) the 3-design is UO for the estimation of $\boldsymbol{\beta}$.

It is not clear if the UO extends to a partial set of the $\boldsymbol{\beta}_{(i)}$'s or to linear functions of the parameters in $\boldsymbol{\beta}_{(i)}$'s such as the sets of contrasts in the elements of $\boldsymbol{\gamma}_{(i)}$. This would require a detailed examination.

**Case II**: $k_1 = k_2 = \cdots = k_b = k$

This is the case discussed in Bhaumik (1995). In this case, each $\mathbf{C}_{d(i)}$ is singular. Further, $\tau_i$ and $\bar{\gamma}_i$ are completely confounded. The $(v-1) \times (v-1)$ information matrix for parameters $\gamma_{(i)}$ adjusted for $\tau_i$ is seen to be

$$\mathbf{F}_i = \begin{bmatrix} \lambda_{i1} & \delta_{i12} & \dots & \delta_{i1v-1} & \delta_{i1v} \\ \vdots & \vdots & \ddots & \vdots & \vdots \\ \delta_{iv1} & \delta_{iv2} & \dots & \delta_{ivv-1} & \lambda_{iv} \end{bmatrix} - \boldsymbol{\lambda}_i \boldsymbol{\lambda}_i'/r_i, \qquad (7.2.8)$$

In this case we shall say that design $d$ is of maximal rank if rank $(\mathbf{F}_i) = v-2$ for all $i$. Such a design would permit estimation of contrasts in the elements of $\gamma_{(i)}$. It will also permit estimation of contrasts in $\theta_i = \tau_i + (\sum_j \lambda_{ij}\gamma_{i(j)}/r_i)$. We note that this set of contrasts is strongly dependent on the design although all such contrasts are estimable.

Same arguments as in Case I can be used to show that if a 3-design exists, it is UO for the estimation of contrasts in $\gamma_{(i)}$, $i = 1, 2, \ldots, v$. One can also claim UO for the joint estimation of $\boldsymbol{\tau}$ and $\boldsymbol{\gamma}$. However, this result is not very meaningful because the set of estimable parametric functions is design - dependent. This comment does *not* apply if we restrict consideration to BIB designs only because in this case the parameters $\theta_i$ are *not* design dependent.

## 7.3 Split Block Designs

In this section we shall discuss optimality of split block designs. A topic first dealt with by Ozawa *et al.* (2001). Here, the setting consists of a number of sites to be called blocks. For each of the $b$ blocks, the experimental units (eu's) are classified in two ways e.g., in $k_1$ rows and $k_2$ columns. There are two factors $F_1$ and $F_2$ at $v_1$ and $v_2$ levels respectively. The levels of $F_1 (F_2)$ are allocated to the $k_1 (k_2)$ rows (columns) according to a binary incomplete block design. Here, we assume that $k_1 \le v_1 (k_2 \le v_2)$. Such a design is called a split block design. For given $(b, k_1, k_2, v_1, v_2)$ the problem is to choose an appropriate design for optimal estimation of the factorial effects.

Ozawa *et al.* (2001) showed that if a balanced design (to be described later here) exists, it is universally optimal (UO) for the estimation of $F_1F_2$ interaction. Their result is for combined estimation of $F_1F_2$ from three of the four strata (to be described below) using generalized least squares estimation (GLSE) and the class of competing designs is restricted to those designs where the row-treatment and the column-treatment designs are balanced incomplete block designs (BIBD's).

Here, we consider the estimation problem separately for each stratum and show that for each stratum the balanced design, if it exists, is UO. The class of competing designs is different for different strata.

We describe the four strata as follows.

**Stratum 1:** This consists of the $(b-1)$ comparisons among the $b$ block totals. It will be shown that these can provide information on the main effects as well as the interactions when the block effects are assumed to be random.

**Stratum 2(3):** This consists of the $b(k_1-1)$ comparisons among the rows and $b(k_2-1)$ comparisons among the columns in the same block. It will be shown that these provide information on $F_1(F_2)$ and on $F_1F_2$. Here the row (column) effects are regarded as random variables.

**Stratum 4:** This consists of the $b(k_1-1)(k_2-1)$ comparisons for the $b$ blocks which are free from the row and the column effects, regardless of whether these are fixed or random effects. These can be used for estimation of $F_1F_2$ only.

It may be remarked that if the row and column effects are regarded as fixed, there is *no* information available from Strata 2 and 3 on any of the factorial effects $(F_1, F_2, F_1F_2)$.

An advantage of considering the analysis separately in each stratum is that it permits us to carry out exact tests of hypothesis or obtain confidence intervals (under the usual assumption of normality). This is not possible if the analysis is done from all the strata combined as is done by Ozawa *et al.* (2001).

Let $m_p^{il} = 1(0)$ if the $l$th row in the $i$th block receives (does not receive) level $p$ of $F_1$. Similarly, we define $n_q^{is} = 1(0)$ if the $s$th column in the $i$th block receives (does not receive) level $q$ of $F_2$.

The model may now be written as follows. Let $y_{ils}$ denote the observation for the $l$th row and the $s$th column for the $i$th block. We postulate that

$$\begin{aligned} y_{ils} &= \mu_{..} + \delta_i + \alpha_{il} + \beta_{is} + \sum_p m_p^{il}(\mu_{p.} - \mu_{..}) + \sum_q n_q^{is}(\mu_{.q} - \mu_{..}) \\ &\quad + \sum_p \sum_q m_p^{il} n_q^{is}(\mu_{pq} - \mu_{p.} - \mu_{.q} + \mu_{..}) + \xi_{ils}. \end{aligned} \tag{7.3.1}$$

Here, $\delta_i$ is the effect of the $i$th block, $\alpha_{il}$ is the effect of the $l$th row in the $i$th block, $\beta_{is}$ is the effect of the $s$th column in the $i$th block and $\xi_{ils}$ are the observational errors assumed to be uncorrelated with zero means and to have a common variance denoted by $\sigma^2$. Further, $\mu_{pq}$ is the effect of the $(p,q)$ combination of the levels of $(F_1, F_2)$ and $\mu_{p.} = \sum_q \mu_{pq}/v_2$, $\mu_{.q} = \sum_p \mu_{pq}/v_1$, $\mu_{..} = \sum\sum \mu_{pq}/v_1v_2$ denote the means as indicated. The effects $\delta_i$, $\alpha_{il}$ and $\beta_{is}$ may be fixed or random. We shall clearly specify the nature of these effects for each stratum. A design is specified by $(m_p^{il}, n_q^{is}; i = 1, 2, \ldots, b; l = 1, 2, \ldots, k_1; s = 1, 2, \ldots, k_2)$. The class of designs with given values of $(b, k_1, k_2, v_1, v_2)$ will be denoted by $\mathcal{D}$. It will be assumed that $m_{pi} = \sum_l m_p^{il}$ and $n_{qi} = \sum_s n_q^{is}$ take values 0 or 1. This will be called binary assumption.

For the analysis in the bottom stratum (Stratum 4), we regard the $\delta_i$'s, $\alpha_{il}$'s and $\beta_{is}$'s fixed. Here, $\sigma_4^2 = \sigma^2$. For Stratum 2, we regard $\mu$ and $\delta_i$'s fixed whereas $\alpha_{il}$'s are uncorrelated random variables (uncorrelated also with $\xi_{ils}$'s) with zero means and a common variance to be denoted by $\sigma_r^2$. It is easy to see that $\sigma_2^2 = \sigma^2 + k_2\sigma_r^2$. Similar remarks apply to Stratum 3 giving $\sigma_3^2 = \sigma^2 + k_1\sigma_c^2$. Finally, in Stratum 1, $\delta_i$'s are also assumed to be uncorrelated random variables (uncorrelated with the $\alpha_{il}$'s, $\beta_{is}$'s and $\xi_{ils}$'s) with zero means and having a common variance $\sigma_b^2$. This gives $\sigma_1^2 = \sigma^2 + k_2\sigma_r^2 + k_1\sigma_c^2 + k_1k_2\sigma_b^2$. The observational contrasts belonging to different strata are easily seen to be uncorrelated.

The incidence matrices for the row and column designs are given by $\mathbf{M} = (m_{pi})$ and $\mathbf{N} = (n_{qi})$ respectively. These determine the final design excepting the pairing of the blocks of the two designs. We shall assume three stage randomization consisting of the randomization over the blocks and the randomization over the rows and the columns. The replication vectors are given by

$$\mathbf{r}_{(1)} = (r_{11}, \ldots, r_{1v_1})' \text{ and } \mathbf{r}_{(2)} = (r_{21}, \ldots, r_{2v_2})',$$

respectively. Here, $\mathbf{r}_{(1)} = \mathbf{M1}$ and $\mathbf{r}_{(2)} = \mathbf{N1}$. The concurrences are given by $\lambda^1(p,p') = \sum_i m_{pi}m_{p'i}$ and $\lambda^2(q,q') = \sum_i n_{qi}n_{q'i}$ respectively. Clearly, $\lambda^1(p,p) = r_{1p}$ and $\lambda^2(q,q) = r_{2q}$. Following Ozawa *et al.* (2001) we use the following notation (somewhat modified in some places):

| | |
|---|---|
| $\lambda(p,p';q,q')$ | # of of blocks in which levels $p$ and $p'$ of $F_1$ occur *and* levels $q$ and $q'$ of $F_2$ occur ($p \neq p'; q \neq q'$), |
| $\lambda(p,p';q)$ | # of of blocks in which levels $p$ and $p'$ of $F_1$ occur *and* level $q$ of $F_2$ occurs ($p \neq p'$), |
| $\lambda(p;q,q')$ | # of of blocks in which level $p$ of $F_1$ occurs *and* levels $q$ and $q'$ of $F_2$ occur, |
| $\lambda(p;q)$ | # of blocks in which level $p$ of $F_1$ occurs *and* level $q$ of $F_2$ occurs. |

A design for which $\lambda(p,p';q,q')$ is the same for all $p \neq p'$ and $q \neq q'$ is termed balanced incomplete split block design (BISBD) by Ozawa *et al.* (2001). They have also shown that for such a design, $\lambda(p,p';q)$, $\lambda(p;q,q')$ and $\lambda(p;q)$ are all constants. Further, the two block designs are also balanced incomplete block designs (BIBD's) with parameters $(b,k_1,v_1,r_1,\lambda_1)$ and $(b,k_2,v_2,r_2,\lambda_2)$ respectively. Here $r_1$ and $r_2$ are replication numbers for the two designs, whereas $\lambda_1$ and $\lambda_2$ are the two concurrence numbers.

For the analysis of Stratum 4, we eliminate the row and column effects in each block. When we do this, the main effects of the two factors are also eliminated. Routine analysis shows that the trace of the information matrix for $\gamma_{pq} = \mu_{pq} - \mu_{p,} - \mu_{.q} + \mu_{..}$ is $b(k_1-1)(k_2-1)$. This is the same for *all* binary designs.

It is easy to verify that for any design $d$ with information matrix $\mathbf{C}$, $\sum_g \mathbf{C}_g$ has the same structure as the $\mathbf{C}$-matrix for a BISBD i.e. these are proportional. Further, since the trace is the same for all designs, it follows that $|\mathcal{G}|\mathbf{C}^* - \sum_g \mathbf{C}_g = 0$ where $\mathbf{C}^*$ is the information matrix for the BISBD. Here, the summation is over all permutations of the $v_1$ levels of $F_1$ and also over the $v_2$ levels of $F_2$ so that $\mathcal{G}$ denotes the product group of these two permutation groups. Thus, by Theorem 7.1.1, if a BISBD exists, it is UO among *all* designs for which $\mathbf{M}$ and $\mathbf{N}$ are binary. This is in contrast with the result of Ozawa *et al.* who required that the row and column designs are both balanced. This restriction on the class of competing designs is not needed for the UO of the design to hold.

For estimation in Stratum 2 we regard the row-effects as random variables and write down the model for the row sums. When we eliminate the $\delta_i$'s, the column effects and the main effects of $F_2$ are eliminated. It turns out that the main effects of $F_1$ viz. $\mu_{p.}$'s are easily eliminated *only* if we assume that the row design is balanced. We shall now make this assumption.

It is shown in Shah (2000) that if the class of competing designs, *i.e.* a sub-class of $\mathcal{D}$ for which the row designs are balanced contains a BISBD, the BISBD has maximum trace for $\mathbf{C}$. Further, for any competing design, $\sum_g \mathbf{C}_g$ is proportional to $\mathbf{C}^*$, the information matrix for the BISBD. Since $\mathbf{C}^*$ has maximum trace, it follows that $|\mathcal{F}|\mathbf{C}^* - \sum_g \mathbf{C}_g$ is *nnd* and hence the BISBD is UO for the estimation of $F_1F_2$ for the subclass under consideration.

It is reasonable to conjecture that the BISBD is also UO for the estimation of $F_1$ from Stratum 2. However, we do not pursue this at this time. Similar results hold for the estimation of $F_1F_2$ from Stratum 3.

For estimation from Statum 1, we regard row effects, column effects and block effects random and only eliminate the general mean. For $B_i = \sum_l \sum_s y_{ils}$ our model is

$$B_i = k_1k_2\mu_{..} + \sum_p m_{pi}k_2(\mu_{p.} - \mu_{..}) + \sum_p n_{qi}k_1(\mu_{.q} - \mu_{..}) + \sum_p \sum_q m_{pi}n_{qi}\gamma_{pq} + \eta_i \qquad (7.3.2)$$

where, as discussed earlier, $V(\eta_i) = \sigma_1^2 + k_2\sigma_r^2 + k_1\sigma_c^2 + k_1k_2\sigma_b^2$. Elements of the coefficients matrix for $\mu_{p.} - \mu_{..}$ eliminating $\mu_{..}$ are seen to be $k_2^2r_{1p}(1-r_{1p}/b)$ diagonally and $k_2^2(\lambda^1(p,p')-r_{1p}r_{1p'}/b)$ off-diagonally. Thus, for elimination of $\mu_{p.}$, we need the row-design to be balanced. Similarly, we need the column-design also to be balanced. The $\mu_{p.} - \mu_{..}, \mu_{.q} - \mu_{..}$ part eliminating the general mean is seen to have coefficients $e\lambda(p;q) - f$ where $e$ and $f$ are appropriate constants. Thus, elimination of $\mu_{p.}$'s would be much easier if $\lambda(p;q)$ is constant. For this case, it turns out that $e\lambda(p;q) - f = 0$.

We shall now assume that the row and the column designs are balanced *and* that $\lambda(p;q) = bk_1k_2/v_1v_2 = \theta$, say. It is shown in Shah (2000) that if the class of such designs contains a BISBD, the information matrix for the $\gamma_{pq}$'s has maximum trace for this design. Again by symmetry considerations it follows that for any design $d$, $\sum_g \mathbf{C}_g$ is proportional to $\mathbf{C}^*$, the information matrix for the BISBD. Since $\mathbf{C}^*$ has maximum trace, it follows that $|\mathcal{F}|\mathbf{C}^* - \sum_g \mathbf{C}_g$ is *n.n.d.* and hence the BISBD is UO.

Thus, we have UO for the estimation of $F_1F_2$ in *each* stratum. For Stratum 4, the class of competing designs is the class of *all* binary designs, for Stratum 2(3) we additionally require that the row (column) design is balanced whereas for Stratum 1 we require both the designs to be balanced *and* equi-replicate.

Ozawa *et al.* (2001) regarded $\delta_i$'s as fixed rather than random. When one eliminates the $\delta_i$'s one gets only the *within* blocks contrasts. Consequently, one does *not* get any information from Stratum 1.

For a BISBD the **C**-matrix for the estimation of $\gamma$'s has the same form for each stratum. This form is

$$\mathbf{C} = \begin{pmatrix} \mathbf{A} & \mathbf{B} & \dots & \mathbf{B} \\ \mathbf{B} & \mathbf{A} & \dots & \mathbf{B} \\ \vdots & \vdots & \ddots & \vdots \\ \mathbf{B} & \mathbf{B} & \dots & \mathbf{A} \end{pmatrix},$$

where each of **A** and **B** is a completely symmetric (c.s.) matrix of order $v_2 \times v_2$. Since $\sum_q \gamma_{pq} = 0$ and $\sum_p \gamma_{pq} = 0$, it is easily seen that if $\mathbf{A} = a\mathbf{I} + b\mathbf{J}$ and $\mathbf{B} = c\mathbf{I} + d\mathbf{J}$, a generalized inverse for **C** is given by $(a-c)^{-1}\mathbf{I}$. The values of $(a-c)$ for each stratum are given in the following Table, where $\alpha = b/v_1v_2(v_1-1)(v_2-1))$.

**Table 7.1.** The values of $a - c$ by stratum

| Stratum | $a - c$ |
|---|---|
| 1 | $\alpha(v_1 - k_1)(v_2 - k_2)$ |
| 2 | $\alpha v_1(k_1 - 1)(v_2 - k_2)$ |
| 3 | $\alpha v_2(k_2 - 1)(v_1 - k_1)$ |
| 4 | $\alpha v_1 v_2(k_1 - 1)(k_2 - 1)$ |

We may regard $a - c$ as a measure of information provided by the stratum. The above shows that the value of $a - c$ for Stratum 4 is considerably larger than that for Stratum 1. For a BISBD with $b = 16$, $v_1 = v_2 = 4$, $k_1 = k_2 = 3$, $a - c$ for Stratum 4 is 64 times the value for Stratum 1. An example of such a BISBD is obtained by taking a direct product of a BIBD having parameters $b = v = 4$, $r = k = 3$, $\lambda = 1$ with itself.

Thus, Stratum 4 provides a great deal more information on $\gamma$ than Stratum 1. Further, $\sigma_1^2$ is considerably larger than $\sigma_4^2$. We note that Stratum 4 is the most informative one. For this Stratum, the BISBD is UO in the class of *all* binary designs. For other strata the optimality is within restricted classes. However, these strata are less informative. For Stratum 1, which is the least informative, the optimality is within the most restricted class!

The following table indicates information on the estimability of different effects from different strata analyses. It is clear from Table 7.2 that the estimates of $F_1$, $F_2$, $F_1F_2$, using independent information from different strata can be obtained. Procedures for obtaining combined estimates with desirable properties are available in Shah (2000).

**Table 7.2.** Estimability ($\surd$) of the effects $F_1$, $F_2$ and $F_1F_2$ by stratum

| | Effects | | |
|---|---|---|---|
| Stratum | $F_1$ | $F_2$ | $F_1F_2$ |
| 1 | $\surd$ | $\surd$ | $\surd$ |
| 2 | $\surd$ | – | $\surd$ |
| 3 | – | $\surd$ | $\surd$ |
| 4 | – | – | $\surd$ |

## 7.4 Nested Experimental Designs

### 7.4.1 Introduction

In a variety of situations in agricultural, industrial or even environmental applications of experimental designs, we might find some "blocking" factors forming a "nested" pattern in the following sense. There are several distinct levels of the first factor, say $A$. Corresponding to each level of $A$, there are again several distinct levels of the second factor, say $B$; and so on. We say that $A$ is the "nesting" factor while $B$ is the "nested" factor. In a study to examine differential effects of some treatments in yield production , for example, agricultural plots may be selected from different regions and in each region, possibly from different types of land having different soil fertility, irrigation facility and climatic conditions. The treatments are then administered in the experimental plots and the resulting data analyzed. In this case we may consider the regions as being the levels of the first factor and the types of land in each region as those of the second factor. This gives rise to a nested design with one nested and one nesting factor. This is known in the literature as "nested block design".

We now formally define a nested block design in its most general set-up. There are two blocking factors $A$ and $B$. $A$ is the nesting factor at "a" levels and corresponding to the $i$th level $A(i)$ of $A$, the nested factor $B$ has $b_i$ levels viz., $B(i1)$, $B(i2)$, ..., $B(ib_i)$. At the location $(ij)$ representing the level combination $[A(i), B(ij)]$, there are $k(ij)$ $[\geq 1]$ experimental units (eu's) so that altogether we have $n = \sum\sum k(ij)$ eu's. Further to this, we have $v$ treatments whose differential effects are to be compared in this design set-up. We may denote the design as $[a, \{(b_i, k(ij))\} | 1 \leq j \leq b_i; 1 \leq i \leq a]$. In the special case: $b_i = b$, $k(ij) = k$ for all $i$, $j$ , we refer to the nested block design as based on a *symmetric layout* with parameters $a, b, k, v$ and denote it by $[a, b, k, v]$. Thus, in the nested block design $[a, b, k, v]$, there are on the whole "$abk$" eu's available in "a" sets of "$bk$" each and within each such set, there are again $b$ blocks of size $k$ each.

Likewise, a nested row-column (nested r-c) design in a symmetric layout with parameters $[a, b, c, v]$ involves "a" levels of the nesting factor $A$. However, this time there are two nested factors : $B$ (with $b$ levels) and $C$ (with $c$ levels) under each level $A(i)$ of $A$. Corresponding to the level combination $\{A(i),(B(ij),C(ik))|1 \leq i \leq a, 1 \leq j \leq b, 1 \leq k \leq c\}$, we have only one experimental unit. A generalization of this set-up is described in Section 7.4.3.

Lattice designs are simplest examples of nested block designs (with $b = k = \sqrt{v}$, $v$ a perfect square) and nested row-column designs (with $b = c = \sqrt{v}$). It may be noted that resolvable BIBD's may serve as examples of nested block designs with levels of the nesting factor acting on each component set of blocks of the nested factor, thereby providing a full replication of the treatments in each set.

Two useful early references are Srivastava (1978) and Singh and Dey (1979). Morgan (1996) provides a detailed account of the possible models, analysis, construction and optimality of designs in a nested set-up. Whereas optimality aspects of nested block designs have been studied for quite sometime, those of nested row-column designs have only been studied recently, starting with the work of Bagchi *et al.* (1990). A detailed study of row-column designs is to be found in Shah and Sinha (1996). The latest reference is Shah and Sinha (2001b) for a review of some aspects of such designs.

### 7.4.2 Nested Block Designs

We first discuss the case of a fixed effects additive model. The vector $\mathbf{Y}$ of recorded observations can be modeled as

$$\mathbf{Y} = \mu\mathbf{1} + \mathbf{L}_1\boldsymbol{\beta}^{(1)} + \mathbf{L}_2\boldsymbol{\beta}^{(2)} + \mathbf{T}\boldsymbol{\tau} + \mathbf{e}. \tag{7.4.1}$$

In (7.4.1), $\boldsymbol{\beta}^{(1)}(a \times 1)[\boldsymbol{\beta}^{(2)}(\sum b_i \times 1)]$ represents the vector of fixed effects of the levels of the nesting factor $A$ (of the nested factor $B$, respectively); $\mathbf{L}_1[n \times a]$ and $\mathbf{L}_2[n \times \sum b_i]$ are the incidence matrices for the levels of $A$ and $B$, respectively; $\mathbf{T}[n \times v]$ is the treatment effects incidence matrix; $\boldsymbol{\tau}$ is the vector of treatment effects and $\mu$ represents the overall mean. The error vector $\mathbf{e}$ is assumed to have zero mean and satisfy the conditions for homoscedasticity. We denote the error variance by $\sigma^2$.

The analysis of data under a ***fixed effects additive model*** as in (7.4.1) has been referred to by Morgan (1996) as ***"bottom analysis"***. For a symmetric layout $[a, b, k, v]$, he deduced that the information matrix (**C**-matrix, as it is called) for the treatment effects is given by (ignoring the multiplier $\sigma^{-2}$)

$$\mathbf{C}(\boldsymbol{\tau}|F) = \mathbf{T}'[\mathbf{I} - \mathbf{P}_2]\mathbf{T}, \tag{7.4.2}$$

where $F$ refers to the model with fixed effects and $\mathbf{P}_2$ denotes the projection matrix on $\mathcal{C}(\mathbf{L}_2)$, the column space of $\mathbf{L}_2$.

It is not difficult to assert that under (7.4.1) and the most general experimental set-up $[a, \{(b_i, k(ij))\} 1 \leq j \leq b_i; 1 \leq i \leq a]$, the same information matrix would be attained for the bottom analysis. In other words, the nesting factor ($A$) has no role to play in the fixed effects case even in a general nested design setting. This is because $\mathcal{C}(\mathbf{L}_2) \supseteq \mathcal{C}(\mathbf{L}_1) \supseteq \mathcal{C}(\mathbf{1})$.

**Remark 7.4.1** Note that implicit in the model is the tacit assumption that the effects of the nested factor within (each level of the nesting factor) as well as between different levels of the nesting factor are all *distinct* . Thus, in effect, we have a total of $\sum b_i$ parameters in $\boldsymbol{\beta}^{(2)}$. The representation in (7.4.2) can then be viewed as follows. For each level of the nesting factor, we have a block design involving all or a subset of the $v$ treatments. We derive an expression for the underlying **C**-matrix of the treatments involved in each and expand it to a $v \times v$ matrix by adding zero's. All such **C**-matrices added together will result into the final **C**-matrix in (7.4.2).

In the nested block design symmetric layout set-up $[a, b, k, v]$, Morgan (1996) also described a *"decomposability"* property of $\mathbf{C}(\boldsymbol{\tau}|R)$ under a *random effects model*, the corresponding analysis being referred to by him as the *"full analysis"*. The relevant model is

$$\mathbf{Y} = \mu\mathbf{1} + \mathbf{L}_1\mathbf{b}^{(1)} + \mathbf{L}_2\mathbf{b}^{(2)} + \mathbf{T}\boldsymbol{\tau} + \mathbf{e} \tag{7.4.3}$$

where the random error vectors $\mathbf{b}^{(1)}$ and $\mathbf{b}^{(2)}$ are uncorrelated with each other as also with $\mathbf{e}$ and further

$$E(\mathbf{b}^{(i)}) = \mathbf{0}, \quad \mathbf{V}(\mathbf{b}^{(i)}) = \sigma_i^2\mathbf{I}, \quad i = 1, 2. \tag{7.4.4}$$

It is readily seen that

$$\mathbf{V}(\mathbf{Y}) = \mathbf{L}_1\mathbf{L}_1'\sigma_1^2 + \mathbf{L}_2\mathbf{L}_2'\sigma_2^2 + \sigma^2\mathbf{I} = \boldsymbol{\Sigma}, \text{ say.} \tag{7.4.5}$$

Morgan (1996) deduced that

$$\begin{aligned} \mathbf{C}(\boldsymbol{\tau}|R) &= \mathbf{T}'[\boldsymbol{\Sigma}^{-1} - \boldsymbol{\Sigma}^{-1}\mathbf{1}(\mathbf{1}'\boldsymbol{\Sigma}^{-1}\mathbf{1})^{-1}\mathbf{1}'\boldsymbol{\Sigma}^{-1}]\mathbf{T} \\ &= a_1^{-1}\mathbf{C}_0(\boldsymbol{\tau}|F) + (a_2^{-1} - a_1^{-1})\mathbf{C}_1(\boldsymbol{\tau}|F) \\ &\quad + (a_3^{-1} - a_2^{-1})\mathbf{C}_2(\boldsymbol{\tau}|F), \end{aligned} \tag{7.4.6}$$

where

$$\begin{aligned} \mathbf{C}_0(\boldsymbol{\tau}|F) &= \mathbf{r}^\delta - \mathbf{r}\mathbf{r}'/n \\ &= \text{C-matrix based on a completely randomized design (CRD)} \end{aligned}$$

$$\begin{aligned} \mathbf{C}_1(\boldsymbol{\tau}|F) &= \mathbf{r}^\delta - \mathbf{N}_A\mathbf{N}_A'/bk \\ &= \text{C-matrix based on a block design with the levels} \\ &\quad \text{of the factor } A \text{ as forming fixed effects} \end{aligned}$$

$$\begin{aligned} \mathbf{C}_2(\boldsymbol{\tau}|F) &= \mathbf{r}^\delta - \mathbf{N}_B\mathbf{N}_B'/k \\ &= \text{C-matrix based on a block design with all levels} \\ &\quad \text{of the factor } B \text{ as forming fixed effects} \end{aligned} \tag{7.4.7}$$

and

$$a_1 = bk\sigma_1^2 + k\sigma_2^2 + \sigma^2; \quad a_2 = k\sigma_2^2 + \sigma^2; \quad a_3 = \sigma^2. \tag{7.4.8}$$

Clearly, expression (7.4.6) refers to the decomposability property.

We may now claim that this sort of property holds even for a *mixed effects model* under the symmetric layout $[a, b, k, v]$ when the nesting factor $A$ is of fixed effects while the nested factor $B$ is of random effects. To see this, note that when $A$ is fixed and $B$ is random, model (7.4.1) changes to

$$\mathbf{Y} = \mu\mathbf{1} + \mathbf{L}_1\boldsymbol{\beta}^{(1)} + \mathbf{L}_2\mathbf{b}^{(2)} + \mathbf{T}\boldsymbol{\tau} + \mathbf{e}, \tag{7.4.9}$$

so that

$$\mathbf{V}(\mathbf{Y}) = \mathbf{L}_2\mathbf{L}_2'\sigma_2^2 + \sigma^2\mathbf{I} = \boldsymbol{\Sigma}^*, \text{ say.} \tag{7.4.10}$$

Thus, the information matrix for the treatment parameters is given by

$$\mathbf{C}(\boldsymbol{\tau}|M) = \mathbf{T}'(\boldsymbol{\Sigma}^{*-1} - \boldsymbol{\Sigma}^{*-1}\mathbf{L}^*(\mathbf{L}^{*\prime}\boldsymbol{\Sigma}^{*-1}\mathbf{L}^*)^-\mathbf{L}^{*\prime}\boldsymbol{\Sigma}^{*-1})\mathbf{T}, \tag{7.4.11}$$

where $\mathbf{L}^* = (\mathbf{1}, \mathbf{L}_1)$.

Following Morgan (1996), we write

$$\begin{aligned}\boldsymbol{\Sigma}^* &= \mathbf{L}_2\mathbf{L}_2'\sigma_2^2 + \sigma^2\mathbf{I} = k\sigma_2^2\mathbf{P}_2 + \sigma^2\mathbf{I} = (a_1^* - a_2^*)\mathbf{P}_1 + (a_2^* - a_3^*)\mathbf{P}_2 + \sigma^2\mathbf{I} \\ &= a_1^*\mathbf{P}_1 + a_2^*(\mathbf{P}_2 - \mathbf{P}_1) + a_3^*(\mathbf{I} - \mathbf{P}_2),\end{aligned} \tag{7.4.12}$$

where

$$\begin{aligned}&\mathbf{P}_1 = \mathbf{L}^*(\mathbf{L}^{*\prime}\mathbf{L}^*)^-\mathbf{L}^{*\prime}, \quad \mathbf{P}_2 = \mathbf{L}_2(\mathbf{L}_2'\mathbf{L}_2)^-\mathbf{L}_2', \\ &a_1^* = a_2^* = k\sigma_2^2 + \sigma^2 \text{ and } a_3^* = \sigma^2.\end{aligned} \tag{7.4.13}$$

Next note that $\mathbf{P}_1 \perp (\mathbf{P}_2 - \mathbf{P}_1) \perp (\mathbf{I} - \mathbf{P}_2)$ so that

$$(\boldsymbol{\Sigma}^*)^{-1} = \frac{1}{a_1^*}\mathbf{P}_1 + \frac{1}{a_2^*}(\mathbf{P}_2 - \mathbf{P}_1) + \frac{1}{a_3^*}(\mathbf{I} - \mathbf{P}_2). \tag{7.4.14}$$

Further,

$$\mathbf{P}_1\mathbf{L}^* = \mathbf{L}^*, \quad (\mathbf{P}_2 - \mathbf{P}_1)\mathbf{L}^* = \mathbf{0} \text{ and } (\mathbf{I} - \mathbf{P}_2)\mathbf{L}^* = \mathbf{0}. \tag{7.4.15}$$

In view of (7.4.11) and (7.4.14), it now follows that

$$\begin{aligned}\mathbf{C}(\boldsymbol{\tau}|M) &= \frac{1}{a_2^*}\mathbf{T}'(\mathbf{I} - \mathbf{P}_1)\mathbf{T} + \frac{1}{a_3^*}\mathbf{T}'(\mathbf{I} - \mathbf{P}_2)\mathbf{T} \\ &= \frac{1}{a_2^*}\mathbf{C}_1(\boldsymbol{\tau}|F) + [\frac{1}{a_3^*} - \frac{1}{a_2^*}]\mathbf{C}_2(\boldsymbol{\tau}|F).\end{aligned} \tag{7.4.16}$$

This establishes that decomposability property holds even under a mixed effects model with the second factor random.

**Remark 7.4.2** The bottom analysis described above holds under the most general form of data even for higher order nested layouts. However, the full analysis holds only when we have symmetric layouts in the sense of $[a, b, c, \ldots, k, v]$.

**Remark 7.4.3** If we consider the situation where the first factor is random and the second factor is fixed, then the decomposability property does not hold even under a symmetric nested layout.

**Remark 7.4.4** The decomposability result can be extended to mixed effects models in higher order nested symmetric layouts. However, it is imperative that the first few nesting factors are treated as fixed and the rest are *all* treated as random.

**Remark 7.4.5** It may be noted that the full analysis requires knowledge of the variance components $\sigma_1^2$, $\sigma_2^2$ and $\sigma^2$. These are generally not known. However, at least for the symmetric case, these can be estimated from the data using standard methods in variance components analysis. Of course, exact properties of the resulting estimators of the treatment comparisons obtained in the full analysis are not easily tractable.

### 7.4.3 Optimal Nested Block Designs

Consider the symmetric layout $[a, b, k, v]$. Note that the information matrix for comparison of the treatment effects (in the bottom analysis) corresponds to that of a block design involving "$ab$" blocks each of size "$k$". Suppose $k < v$. Then, if there exists a BIBD with parameters $B = ab$, $V = v$, $R = abk/v$, $K = k$ and $\lambda$, then it will result into an universally optimal design in the sense of Kiefer (1975). Also if $k > v$, then we would require existence of a balanced block design (BBD). Of course when $k = v$, we always have the randomised block design RBD available. Again note that for the full analysis, the treatment comparisons information matrix is composed of a combination of two such matrices both based on block designs: one with "$a$" blocks each of size "$bk$" and the other as described above viz., one with "$ab$" blocks each of size "$k$". Therefore, it would be desirable to have both parts as BBD / RBD / BIBD, depending on the situations. Non-trivial situation corresponds to [BIBD, BIBD] in which case the resulting nested design is called "nested BIBD" and will be denoted by $N$ [BIBD, BIBD]. Since each of $\mathbf{C}_1(\tau|F)$ and $\mathbf{C}_2(\tau|F)$ corresponds to a UO design, the design is UO for the full analysis as well.

Various methods of construction of $N$ [BIBD, BIBD]'s are available. See Morgan (1996). We will discuss one method below.

**Theorem 7.4.1** *Suppose there exists a BIBD$(B, V, R, K = bk, \Delta)$ and for some $r > k$, also there exists a resolvable BIBD $(br, bk, r, k, \lambda)$. Then we can construct an $N$[BIBD$(Br, V, Rr, K, r\Delta)$, BIBD$(Bbr, V, Rr, k, \Delta\lambda)$].*

**Proof.** The BIBD$(mB, V, mR, K, m\Delta)$ is formed by taking m replications of the BIBD$(B, V, R, K, \Delta)$.

Consider the first block of the BIBD$(B, V, R, K, \Delta)$. There are $K = bk$ treatments in this block. Use these treatments to construct a resolvable BIBD$(br, bk, r, k, \lambda)$. Carry out the same, taking in turn, other blocks of the BIBD$(B, V, R, K, \Delta)$. There are altogether Bbr blocks which collectively form the BIBD$(Bbr, V, Rr, k, \lambda^*)$ . It is now enough to take one component (i.e., a full replication) from each resolvable BIBD and identify the same as a block of the BIBD$(Br, V, Rr, K, r\Delta)$. □

**Corollary 7.4.1** *Whenever s is a prime power, N[BIBD* $(s(s^2-1), s^2, s^2-1, s, s-1)$, *BIBD*$(s(s+1)^2, s^2, (s+1)^2, s, s+1)$*] exists.*

**Proof.** The proof rests on the use of Yates' orthogonal series BIBD $(s^2+s, s^2, s+1, s, 1)$ which is resolvable.

Here is an example of N [BIBD, BIBD] based on the use of Yates' orthogonal series $[OS(1)]$: $a = 60, b = k = 2, v = 16$; $N$[BIBD$(60, 16, 15, 4, 3)$, BIBD$(120, 16, 15, 2, 1)$]. We use $d =$ resolvable BIBD$(6, 4, 3, 2, 1)$ in combination with $D =$ BIBD$(20, 16, 5, 4, 1)$. For the first BIBD, we use 3 replications of $D$. Each block of $D$ is then used in conjunction with d to generate 6 blocks , each of size 2. These are made into three sets for the 3 replications. Thus, for example, $(1, 2, 3, 4)$ of $D$ will actually appear as

$$\begin{matrix} 1\,2 & 13 & 14 \\ 3\,4 & 24 & 23 \end{matrix}$$

in the 3 replications. The rest is clear. □

### 7.4.4 Nested Row-Column Designs

A nested row-column design with a symmetric layout has the design parameters $[a, b, c, v]$. In its most general form, the design layout is given by

$$[\{A(i), B(ij), C(ik)\} | 1 \leq i \leq a, 1 \leq j \leq b(i), 1 \leq k \leq c(i)].$$

For the bottom analysis, Remark 7.4.1 is very relevant and the information matrix for the treatment comparisons is to be obtained as the sum of the component **C**-matrices derived for the usual row-column designs and suitably expanded to $v \times v$ matrices. For the full analysis, however, we have to deal with symmetric layouts.

Nested row-column designs have been extensively discussed in Morgan (1996). Optimality study of nested row-column designs was initiated in Bagchi *et al.* (1990). The literature is fairly recent. We do not discuss the details. See also Shah and Sinha (2001b).

## 7.5 Optimality Status of Incomplete Layout Three-Way Balanced Designs

The key references to this section are Agrawal (1966), Shah and Sinha (1990), Heiligers and Sinha (1995) and Saharay (1996). Other related references are Bagchi and Shah (1989), Hedayat and Raghavarao (1975) and Shah and Sinha (1989, 1996).

The set-up is that of a row-column design (also called a three-way design) and the model is that of fixed additive effects with homoscedastic errors. We say that the row and column classification represents known heterogeneity directions and the treatment effects correspond to the third component of variation. Our primary task is to provide efficient comparisons of the treatment effects. Further to this, we have two sets of differential effects, each within a specific heterogeneity direction, to be compared (or at least eliminated). Technically, these are referred to as row effects and column effects.

A three-way design involving $R$ rows, $C$ columns and $v$ treatments is said to possess a complete (incomplete) layout if the number of experimental units (eus) in the design is equal to (less than, respectively) $RC$. The row-column layout is generally made available to the experimenter who has to design treatment allocation over the eu's. The statistical analysis of data arising out of such a design is fairly straightforward and we refer to Shah and Sinha (1996) for this. There are three (assignable) sources of variation: rows, columns and treatments. These correspond to three classifications and the resulting data is often referred to as three-way classified data in the literature.

A three-way design is said to be *three-way balanced* (or, *totally balanced*) whenever the variances of estimates of normalized effects contrasts are the same for each classification. We readily verify that the simplest example of a three-way balanced design is a Latin square which is known to possess a complete layout involving the same number of rows, columns and treatments.

Incomplete three-way balanced designs are not easy to come across. Agrawal (1966) was the first to provide some series of such designs. Afterwards, Hedayat and Raghavarao (1975) also provided one series of such designs.

Note that a three-way balanced design in an incomplete three-way layout provides complete symmetry of the **C**-matrices involving comparisons of row effects, or the column effects, or the treatment effects. In view of this kind of strong symmetry possessed by these designs, it is reasonable to expect that such designs will be optimal for inference on contrasts of the parameters for each of the three sources of variation: rows, columns and treatments.

However, subsequent studies by Shah and Sinha (1990), Heiligers and Sinha (1995) and Saharay (1996) proved otherwise for most of the cases. Below we present salient features of these studies. Once for all, we use the notations $\mathbf{N}_{rc}$, $\mathbf{N}_{rt}$ and $\mathbf{N}_{ct}$ to denote respectively the row-column, row-treatment and column-treatment incidence matrices of appropriate orders in a given context. Likewise, $\mathbf{C}_r$, $\mathbf{C}_c$ and $\mathbf{C}_t$ will denote the **C**-matrices for row effects comparisons, column effects comparisons and treatment effects comparisons, respectively.

It must be emphasized that in a given situation, we have the eu's available only at certain positions in the row-column setting as dictated by the row-column incidence matrix $\mathbf{N}_{rc}$. For given $\mathbf{N}_{rc}$, the allocation of the treatments to the eu's is to be so chosen that the corresponding incidence matrices $\mathbf{N}_{rt}$ and $\mathbf{N}_{ct}$ are feasible and consistent with the layout suggested by $\mathbf{N}_{rc}$.

### 7.5.1 Three-Way Balanced Designs Based on Agrawal's Method 1

We will first present results related to three-way balanced designs with the parameters $R = C = v = 4t + 3$, $v$ a prime, for which the row-column incidence pattern is given by

$$\begin{aligned} &\mathbf{N} = \mathbf{N}_{rc} = (0,1) \text{ with } \mathbf{N} + \mathbf{N}' = \mathbf{J} - \mathbf{I} \\ \text{and} \quad &\mathbf{N}\mathbf{N}' = (rt)\mathbf{I} + t\mathbf{J},\ r = 2t + 1. \end{aligned} \tag{7.5.1}$$

Method 1 of Agrawal (1966) provides a design for which $\mathbf{N}_{rc} = \mathbf{N}_{rt} = \mathbf{N}_{ct} = \mathbf{N}$ as in (7.5.1).

It follows that for Agrawal's design

$$\mathbf{C}_A = \mathbf{C}_r = \mathbf{C}_c = \mathbf{C}_t = [(2t^2 - 1)/t][\mathbf{I} - \mathbf{J}/v]. \tag{7.5.2}$$

For treatment comparisons, however, the $\mathbf{C}_t$-matrix based on specified $\mathbf{N}_{rc}$ as above and arbitrary incidence matrices $\mathbf{N}_{rt}$ and $\mathbf{N}_{ct}$ is given by

$$\begin{aligned} \mathbf{C}_t &= (2t+1)[\mathbf{I} - \mathbf{J}/(4t+3)] - [(2t+1)/t(4t+3)][\mathbf{N}'_{rt}\mathbf{N}_{rt} + \mathbf{N}'_{ct}\mathbf{N}_{ct}] \\ &\quad + [1/t(4t+3)][\mathbf{N}'_{rt}\mathbf{N}\mathbf{N}_{ct} + \mathbf{N}'_{ct}\mathbf{N}'\mathbf{N}_{rt}] \end{aligned} \tag{7.5.3}$$

At this stage, for the choice $\mathbf{N}_{rt} = \mathbf{N}, \mathbf{C}_t$ simplifies to

$$\begin{aligned} \mathbf{C}_t &= [(2t+1)(4t^2 + 2t + 1)/t(4t+3)]\mathbf{I} - [(2t+1)/t(4t+3)](\mathbf{N}'_{ct}\mathbf{N}_{ct}) \\ &\quad + [(t+1)/t(4t+3)](\mathbf{N}'_{ct} + \mathbf{N}_{ct}) \end{aligned} \tag{7.5.4}$$

Now, we note that the last term in (7.5.4) has a positive coefficient. Agrawal's choice viz., $\mathbf{N}_{ct} = \mathbf{N}$ implies $\mathbf{N}'_{ct} + N_{ct} = \mathbf{J} - \mathbf{I}$ [vide (7.5.1)]. This was the point made by Shah and Sinha (1990). They argued that it is possible to make a choice of $\mathbf{N}_{ct}$ in such a way that $\mathbf{N}'_{ct} + \mathbf{N}_{ct}$ has 2 along the diagonal so that it contributes $2(t+1)/t$ to $\mathrm{tr}(\mathbf{C}_t)$ in excess of what is obtained from Agrawal's choice. Of course, for large values of $t$, this increase may not be substantial. We now specialize to $R = C = v = 7$ and $\mathbf{N}_{rc} = (0\ 1\ 1\ 0\ 1\ 0\ 0)$, a circulant of order 7 with elements 0, 1, 1, 0, 1, 0, 0 in the first row.

Shah and Sinha (1990) succeeded in finding an alternative to Agrawal's design for which the $\mathbf{C}_t$-matrix $\mathbf{C}_{S-S}$ say, satisfies the relation $\mathbf{C}_{S-S} > \mathbf{C}_A$ in the Loewner Ordering sense! However, $\mathbf{C}_{S-S}$ turned out *not* to be completely symmetric (c.s.). Their choice is $\mathbf{N}_{ct} = (1\ 0\ 1\ 0\ 0\ 1\ 0)$. Carrying this investigation further, Heiligers and Sinha (1995) re-established superiority of another c.s. $\mathbf{C}_t$-matrix in the same set-up. Below we display all the three competing designs in terms of the row versus treatment incidence matrices (with rows, columns and treatments denoted by the numbers 0, 1, ..., 6 (mod 7):

Agrawal's Design: Initial Row = $\mathbf{N}_{rt}$ = ( - 2 4 - 1 - - ) and other rows are to be obtained by *developing* the initial row.

Shah and Sinha Design: Initial Row = $\mathbf{N}_{rt}$ = ( - 1 4 - 2 - - ) and other rows are to be obtained by *developing* the initial row.

Heiligers and Sinha Design: Initial Row = $\mathbf{N}_{rt}$ = ( - 1 6 - 2 - - ) and other rows are to be obtained by *developing* the initial row.

Computations yield

$$\begin{aligned}\mathbf{C}_A &= (6-1-1-1-1-1-1)/7;\\ \mathbf{C}_{S-S} &= (100-2-3-3-20)/7\end{aligned}$$

and

$$\mathbf{C}_{H-S} = (12-2-2-2-2-2-2)/7 = 2\mathbf{C}_A.$$

It is interesting to note that Shah and Sinha design also dominates over Agrawal's design with respect to row effects comparisons and column effects comparisons as well.

### 7.5.2 Three-Way Balanced Designs Based on Agrawal's Method 3

In Method 3, Agrawal has constructed three-way balanced designs with the parameters $R = C = v$ and $\mathbf{N}_{rc} = \mathbf{N}_{rt} = \mathbf{N}_{ct} = \mathbf{J} - \mathbf{I}$. A simpler method to construct such designs would be to construct Latin squares of order $v \times v$ with diagonal elements all different and then to delete the diagonal. Anyway, such designs have so much symmetry that it seems impossible to beat them for any of the three sources of variation.

Computations yield:

$$\mathbf{C}_A = \mathbf{C}_r = \mathbf{C}_c = \mathbf{C}_t = [v(v-3)/(v-2)][\mathbf{I} - \mathbf{J}/v]. \tag{7.5.5}$$

Further, for arbitrary choices of $\mathbf{N}_{rt}$ and $\mathbf{N}_{ct}$ subject to $\mathbf{N}_{rc} = \mathbf{J} - \mathbf{I}$, it follows that

$$\begin{aligned}\mathbf{C}_t &= \mathbf{r}^d + [1/(v-1)(v-2)]\mathbf{r}\mathbf{r}'\\ &\quad - \frac{1}{v(v-2)}[(v-1)(\mathbf{N}'_{ct}\mathbf{N}_{ct} + \mathbf{N}'_{rt}\mathbf{N}_{rt}) + (\mathbf{N}'_{ct}\mathbf{N}_{rt} + \mathbf{N}'_{rt}\mathbf{N}_{ct})]\,. \end{aligned} \tag{7.5.6}$$

At this stage, Shah and Sinha (1990) argued that Agrawal's design has minimum trace($\mathbf{C}_t$) in the binary class and that it is possible to increase trace($\mathbf{C}_t$) by allowing for unequal replications.

Carrying this investigation further, Saharay (1996) made the choice: $\mathbf{N}_{rt} = \mathbf{J} - \mathbf{I}$; $\mathbf{N}_{ct} = \mathbf{J} - \mathbf{P}$, $\mathbf{P}$ being a cyclic permutation matrix of order $v$, different from the identity permutation. For this choice

$$\mathbf{C}_S = \mathbf{C}_A + [1/v(v-2)][2\mathbf{I} - \mathbf{P} - \mathbf{P}']. \tag{7.5.7}$$

Since $2\mathbf{I}-\mathbf{P}-\mathbf{P}'$ is an *nnd* matrix, we have Loewner domination of $\mathbf{C}_A$ by $\mathbf{C}_S$. Next, she established that designs with $\mathbf{C}$-matrix of the form (7.5.7) are E-optimal. Also, by extensive computer search, she verified that such designs are $A$- and $D$- optimal at least for values of $v$ up to and including 20 within the class of binary equireplicate designs. Finally, she developed a systematic method of constructing such designs.

We now display below the designs by Agrawal, Shah and Sinha and Saharay for the case of $v = 7$. The rows, columns and treatments are denoted by the numbers $1, 2, \ldots, 7$.

Agrawal's Design:

| | | | | | | |
|---|---|---|---|---|---|---|
| – | 2 | 3 | 4 | 5 | 6 | 7 |
| 6 | – | 1 | 2 | 3 | 4 | 5 |
| 4 | 5 | – | 7 | 1 | 2 | 3 |
| 2 | 3 | 4 | – | 6 | 7 | 1 |
| 7 | 1 | 2 | 3 | – | 5 | 6 |
| 5 | 6 | 7 | 1 | 2 | – | 4 |
| 3 | 4 | 5 | 6 | 7 | 1 | – |

Shah and Sinha Design:

| | | | | | | |
|---|---|---|---|---|---|---|
| – | 7 | 6 | 2 | 1 | 3 | 5 |
| 4 | – | 7 | 6 | 2 | 1 | 3 |
| 5 | 4 | – | 7 | 6 | 2 | 1 |
| 3 | 5 | 4 | – | 7 | 6 | 2 |
| 1 | 3 | 5 | 4 | – | 7 | 6 |
| 2 | 1 | 3 | 5 | 4 | – | 7 |
| 7 | 2 | 1 | 3 | 5 | 4 | – |

Saharay Design:

| | | | | | | |
|---|---|---|---|---|---|---|
| – | 1 | 4 | 3 | 2 | 5 | 6 |
| 2 | – | 3 | 4 | 5 | 6 | 7 |
| 3 | 4 | – | 5 | 6 | 7 | 1 |
| 4 | 5 | 6 | – | 7 | 1 | 2 |
| 5 | 6 | 7 | 1 | – | 2 | 3 |
| 7 | 3 | 2 | 6 | 1 | – | 4 |
| 1 | 2 | 5 | 7 | 4 | 3 | – |

It is also interesting to note that in the above example, Shah and Sinha design improves over Agrawal's design with respect to D-optimality criterion only for treatment effects comparisons while for row effects and column effects comparisons, it does improve over Agrawal's with respect to both A- and D-optimality criteria. There is no Loewner domination of Shah and Sinha designs over Agrawal's.

### 7.5.3 Three-Way Balanced Designs Based on Agrawal's Method 4

In his Method 4, Agrawal (1966) constructed designs with the parameters

$$R = C = 4t + 3 \text{ (a prime)}, \; v = 2(4t+3),$$

$$\mathbf{N}_{rc} = \mathbf{J} - \mathbf{I},\ \mathbf{N}_{rt} = (\mathbf{N},\mathbf{N}) \text{ and } \mathbf{N}_{ct} = (\mathbf{N},\mathbf{N})$$
$$\text{with } \mathbf{N} + \mathbf{N}' = \mathbf{J} - \mathbf{I} \text{ and } \mathbf{N}'\mathbf{N} = (t+1)\mathbf{I} + t\mathbf{J}. \qquad (7.5.8)$$

It can be seen that this solution is obtained by duplicating the solution in Agrawal's Method 1. We have already observed that designs based on Method 1 can be improved. This prompted Shah and Sinha (1990) to suggest improved designs in this situation as well. They started with the choice $\mathbf{N}_{rt} = [\mathbf{N}\ \mathbf{P}]$, $\mathbf{N}_{ct} = [\mathbf{P}\ \mathbf{N}]$ for some $\mathbf{P}$ where $\mathbf{N}$ is as in (7.5.8). Then the $\mathbf{C}_t$ matrix assumes the form (apart from a term involving $\mathbf{J}$),

$$\mathbf{C}_t = \begin{pmatrix} (2t+1)\mathbf{I} - \frac{\mathbf{N}'\mathbf{N}}{(4t+2)} & -\frac{\mathbf{N}'\mathbf{P}}{(4t+2)} \\ & (2t+1)\mathbf{I} - \frac{\mathbf{P}'\mathbf{P}}{(4t+2)} \end{pmatrix} - \frac{(4t+2)}{(4t+1)(4t+3)}\mathbf{H}, \qquad (7.5.9)$$

where

$$\mathbf{H} = \begin{pmatrix} \mathbf{P}'\mathbf{P} + \frac{(\mathbf{P}'\mathbf{N}+\mathbf{N}'\mathbf{P})}{(4t+2)} + \frac{\mathbf{N}'\mathbf{N}}{(4t+2)^2} & \mathbf{P}'\mathbf{N} + \frac{(\mathbf{P}'\mathbf{P}+\mathbf{N}'\mathbf{N})}{(4t+2)} + \frac{\mathbf{N}'\mathbf{P}}{(4t+2)^2} \\ & \mathbf{N}'\mathbf{N} + \frac{(\mathbf{N}'\mathbf{P}+\mathbf{P}'\mathbf{N})}{(4t+2)} + \frac{\mathbf{P}'\mathbf{P}}{(4t+2)^2} \end{pmatrix}. \qquad (7.5.10)$$

If we restrict to a binary equireplicate design, then $\mathrm{tr}(\mathbf{P}'\mathbf{P})$ coincides with $\mathrm{tr}(\mathbf{N}'\mathbf{N})$. From (7.5.8) it is evident that in order to maximize $\mathrm{tr}(\mathbf{C}_t)$, we have to make a choice of $\mathbf{P}$ so as to minimize $\mathrm{tr}(\mathbf{P}'\mathbf{N} + \mathbf{N}'\mathbf{P})$. For Agarwal's choice $\mathbf{P} = \mathbf{N}$, we have $\mathrm{tr}(\mathbf{P}'\mathbf{N} + \mathbf{N}'\mathbf{P}) = (2t+1)v$.

At this stage Shah and Sinha (1990) argued that it is indeed possible to make a choice of $\mathbf{P}$ so that $\mathrm{tr}(\mathbf{P}'\mathbf{N} + \mathbf{N}'\mathbf{P}) << v(2t+1)$. It is true for $v = 7$ as was illustrated by them by taking $\mathbf{P} = (1\ 0\ 1\ 0\ 0\ 1\ 0)$, a circulant of order 7. It follows that the resulting design does better than Agrawal's design with respect to A-, D- and E-optimality criteria for treatment effects comparisons. Further, it shows Loewner dominance over Agrawal's for both row effects and column effects comparisons.

### 7.5.4 Concluding Remarks

For Method 2 of Agrawal (1966), it has been conjectured in Shah and Sinha (1990) that Agrawal's designs are at least A- , D- and E-optimal. Optimality of Agrawal's designs given by Method 5 was discussed in Bagchi and Shah (1989) and also in Shah and Sinha (1989). These designs are found to be $\Psi_f$-optimal in the class of equireplicate designs and E- optimal in the unrestricted class. These designs are based on $\mathbf{N}_{rc} = \mathbf{J}$ so that the row-column classification to start with is orthogonal.

# 7.6 Optimal Designs Under Heteroscedastic Errors in Linear Regression

Most optimality studies in the context of linear, quadratic and polynomial regression models deal with homoscedastic error structure. However, some recent studies also reflect situations wherein the use of heteroscedastic error structure in the context of linear regression is called for. See, for example, Abdelbasit and Plackett (1983), Minkin (1987, 1993), Wu (1988), Sitter and Wu (1993), Khan and Yazdi (1988), Ford *et al.* (1992), Atkinson and Cook (1995), Hedayat *et al.* (1997), Sebastiani and Settimi (1997) and Mathew and Sinha (2001).

The key reference to this section is Das *et al.* (2000) and Minkin (1993) and we intend to re-visit an interesting optimality result reported therein. It turns out that there is a unified approach to the problem studied there. We exploit de la Garza phenomenon and Loewner order domination (LOD) studied in Chapter 3 for asymmetric experimental domains and succeed in characterizing a complete class of experiments, using continuous design theory. This study enables us extend optimality results in Minkin (1993) in a natural manner.

## 7.6.1 A Model with Heteroscedastic Errors

Following Minkin (1993), we consider a Poisson count model

$$\begin{aligned}&Y_x \sim \text{Poisson}[\mu(x)]\\ &\text{with } \mu(x) = c(x)e^{\theta(x)},\ \theta(x) = \alpha + \beta x \text{ and } \beta = -\beta^* < 0,\end{aligned} \tag{7.6.1}$$

where $\alpha$ and $\beta$ are unknown parameters and $x$ is a non-stochastic covariate over the domain $\mathcal{T} = [0, \infty)$. Further, $c(x)$ is a known positive quantity, independent of the unknown parameters.

Since we are primarily interested in the regression parameters, we assume,for simplicity, that we are in a situation where we can use approximate theory and accordingly, confine to the n-point design (vide Minkin 1993):

$$d_n = [(0 \le x_1 < \cdots < x_n < \infty;\ 0 < p_1, p_2, \ldots, p_n;\ \sum c(x_i)p_i = 1]. \tag{7.6.2}$$

Here $c(x)$'s are known positive (finite) real numbers.

Next we write down the log-likelihood function and hence deduce the form of the asymptotic information matrix for the parameters as

$$\mathcal{I}(\alpha, \beta) = \sum c(x_i)p_i e^{\theta(x_i)}(1, x_i)'(1, x_i). \tag{7.6.3}$$

Note that the information matrix has an equivalent representation given by

$$\begin{aligned}&\mathcal{I}(\alpha, \beta) = \sum P_i e^{\theta(x_i)}(1, x_i)'(1, x_i);\\ &P_i = c(x_i)p_i,\ 1 \le i \le n;\quad \sum P_i = 1.\end{aligned} \tag{7.6.4}$$

With reference to $\mathcal{I}$, optimality problem refers to optimal choice of the $x_i$'s and the corresponding $P_i$'s subject to $\sum P_i = 1$. Note that Minkin (1993) characterized the nature of optimum design for minimizing $V(\hat{\beta}^{-1})$. As is typical in nonlinear settings, the optimum design depends on the unknown parameters $\alpha$ and $\beta$.

The components of $\mathcal{I}$ are given by

$$I_{11} = \sum_{i=1}^{n} P_i e^{\alpha-\beta^* x_i}, \ I_{12} = \sum_{i=1}^{n} P_i x_i e^{\alpha-\beta^* x_i}, \ I_{22} = \sum_{i=1}^{n} P_i x_i^2 e^{\alpha-\beta^* x_i}. \tag{7.6.5}$$

It now follows that without any loss of generality, we can ignore the factor $e^{\alpha}$ in the expression for $\mathcal{I}$. Further, let us write $\beta^* x_i = x_i^*$ for each $i$. For convenience, we re-name the information matrix as $\mathcal{I}^*$ having elements

$$I_{11}^* = \sum_{i=1}^{n} P_i e^{-x_i^*}, \ I_{12}^* = \sum_{i=1}^{n} P_i x_i e^{-x_i^*}, \ I_{22}^* = \sum_{i=1}^{n} P_i x_i^2 e^{-x_i^*}. \tag{7.6.6}$$

**Remark 7.6.1** We note in passing that $\mathcal{I}^*$ can be written as a convex combination of component $\mathcal{I}^*$'s based on subsets of points in $\mathcal{T}$. In other words,

$$\mathcal{I}^* = \sum_{r=1}^{k} \pi_r \mathcal{I}^*(r) \tag{7.6.7}$$

where $\pi_1, \pi_2, \ldots, \pi_k$ are sums over mutually exclusive and exhaustive subsets of the set of $P_i$'s and the $\mathcal{I}^*(r)$'s are the component information matrices based on the corresponding mutually exclusive and exhaustive subsets of the set of $x_i$'s with the revised $P$-values as $P_i^* = \frac{P_i}{\pi_r}$ whenever $P_i$ is in the $r$th subset of the $P$-values. Minkin (1993) established that for optimal estimation of the slope parameter, an optimal design is a 2-point design including the point 0.

## 7.6.2 Complete Class of Designs in Minkin's Set-Up

We provide here a unified approach to arrive at a very general result to the effect that for inference on the regression parameters, a complete class of continuous designs must necessarily comprise of 2-point designs including the point 0.

Towards establishing the complete class result stated above, first we start with a 2-point design viz.,

$$d_2 = [a, b; P, Q];$$
$$\text{with } 0 < a < b < \infty; \quad 0 < P, Q < 1; \quad P + Q = 1. \tag{7.6.8}$$

Then, writing $a^* = a\beta^*$ and $b^* = b\beta^*$,

$$I_{11}^* = Pe^{-a^*} + Qe^{-b^*}; \tag{7.6.9}$$

$$I_{12}^* = Pae^{-a^*} + Qbe^{-b^*}; \tag{7.6.10}$$

$$I_{22}^* = Pa^2e^{-a^*} + Qb^2e^{-b^*}. \tag{7.6.11}$$

The following theorem is the main result of this subsection.

**Theorem 7.6.1** *Given $d_2$ in (7.6.8), there exists another 2-point design $d_2^* = [0, c; \lambda, 1-\lambda]$ that dominates $d_2$ in the Loewner domination sense i.e.*

$$d_2^* = [0, c; \lambda, 1-\lambda] \succ d_2 = [a, b; P, Q] \tag{7.6.12}$$

*such that $\mathcal{I}^*(d_2^*) - \mathcal{I}^*(d_2)$ is a nnd matrix.*

**Proof.** Utilization of (7.6.6) yields

$$I_{11}^*(d_2^*) - I_{11}^*(d_2) = \lambda + (1-\lambda)e^{-c^*} - Pe^{-a^*} - Qe^{-b^*}; \tag{7.6.13}$$

$$I_{12}^*(d_2^*) - I_{12}^*(d_2) = (1-\lambda)ce^{-c^*} - Pae^{-a^*} - Qbe^{-b^*}; \tag{7.6.14}$$

$$I_{22}^*(d_2^*) - I_{22}^*(d_2) = (1-\lambda)c^2e^{-c^*} - Pa^2e^{-a^*} - Qb^2e^{-b^*}. \tag{7.6.15}$$

We now proceed as in Mandal *et al.* (2000), taking the clue from Pukelsheim (1993). In other words, in the expression for the difference of the two information matrices, we equate both the terms in $(1,1)$th and $(1,2)$th positions to 0 and solve for $c$ and $\lambda$. Then we show that the term in the $(2,2)$th position is strictly positive.

The equations in terms of $c^* = c\beta^*$ and $\lambda$ are given by

$$\lambda + (1-\lambda)e^{-c^*} = Pe^{-a^*} + Qe^{-b^*}; \tag{7.6.16}$$

$$(1-\lambda)ce^{-c^*} = Pae^{-a^*} + Qbe^{-b^*} \tag{7.6.17}$$

whence, eliminating $\lambda$, we obtain the following equation involving $c^*$:

$$\phi(c^*) = w\phi(a^*) + (1-w)\phi(b^*),$$

$$\text{where} \quad \phi(x^*) = \frac{e^{x^*} - 1}{x^*} \text{ and } w = \frac{Pa^*e^{-a^*}}{Pa^*e^{-a^*} + Qb^*e^{-b^*}}. \tag{7.6.18}$$

It is readily seen that the function $\phi(x^*)$ is convex and increasing in $x^*$ over $[0, \infty)$ so that we have a unique solution for $c^*$ above. Moreover, $a^* < c^* < b^*$ i.e., $a < c < b$. Once $c^*$ is known, $\lambda$ is obtained from the relation

$$1 - \lambda = \frac{P(1-e^{-a^*}) + Q(1-e^{-b^*})}{(1-e^{-c^*})}. \tag{7.6.19}$$

It now remains to verify that $0 < \lambda < 1$ which is equivalent to verifying that

$$e^{-c^*} < Pe^{-a^*} + Qe^{-b^*}. \tag{7.6.20}$$

The proof of this claim is given in the Appendix.

Finally, to show strict positivity of the (2, 2)th term in (7.6.15), upon simplification, we find that we have to establish the inequality: $c^* > wa^* + (1-w)b^*$. This follows readily from the strict convexity of the function $\phi(x^*)$ and the defining equation for $c^*$. This establishes the Theorem. □

**Corollary 7.6.1** *Given $d_n$ in (7.6.2), there exists a 2-point design $d_2 = [0, c; \lambda, 1-\lambda]$ that dominates $d_n$ in the Loewner domination sense.*

**Proof.** In view of Remark 7.6.1, the proof follows by induction on the number $n$ of support points of $d_n$. □

We have thus arrived at a characterization of a complete class of continuous designs for inference on the regression parameters in the set-up described in (7.6.1) and (7.6.2). It is now a routine task to determine specific optimal designs for one or both the parameters with respect to various optimality criteria. This is studied in the next section. We note that the asymptotic variance-covariance matrix of the maximum likelihood estimators of $\alpha$ and $\beta$ is obtained by inverting the information matrix $\mathcal{I}$.

### 7.6.3 Specific Optimal Designs

For inference on the parameter $\beta$ or on $\beta^{-1}$ (as discussed in Minkin 1993) with minimum asymptotic variance i.e., maximum information, we are supposed to maximize $I^*_{22.1}$ for proper choices of $c^*$ and $\lambda$.

Algebraically, we have to maximize

$$f(s,c) = sc^2e^{-c^*} - \frac{[sce^{-c^*}]^2}{[1-s(1-e^{-c^*})]},$$

where $s = (1-\lambda)$. Note that this is the same as maximizing

$$f(s,c^*) = sc^{*2}e^{-c^*} - \frac{[sc^*e^{-c^*}]^2}{[1-s(1-e^{-c^*})]}.$$

For fixed $c^*$, we can maximize the above expression in terms of $s$ and obtain $s_{opt}(c^*) = \frac{1}{[1+e^{-c^*/2}]}$ whence, upon substituting the expression for $s$, we need to maximize $f(c^*) = [\frac{c^*}{1+e^{c^*/2}}]^2$. It follows readily that $c^*_{opt} = 2.557$; $\lambda_{opt} = 0.218$. Minkin (1993) deduced this result through a different approach.

For inference on $\alpha$, it follows that an optimal design is singular with $P_{[x=0]} = 1$. For D-optimality involving both the parameters $\alpha$ and $\beta$, we have to maximize $f(s,c^*) = s(1-s)c^{*2}e^{-c^*}$ and this yields $c^*_{opt} = 2.0$, $\lambda_{opt} = 0.5$.

For A-optimality, we have to minimize $f(s,c^*) = \frac{A}{s} + \frac{B}{(1-s)}$, where $A = \frac{e^{c^*}}{c^{*2}}$; $B = 1 + \frac{1}{c^{*2}}$. This yields $c^*_{opt} = 2.261$, $\lambda_{opt} = 0.444$. Again, for $E$-optimality, we have to maximize

$$f(s,c^*) = [1 - L_1 s] - [(1-L_2 s)^2 + Qs^2]^{1/2},$$

where $L_1 = 1 - e^{-c^*} - c^{*2}e^{-c^*}$, $L_2 = 1 - e^{-c^*} + c^{*2}e^{-c^*}$ and $L = 4c^{*2}e^{-2c^*}$.

Numerical calculations yield: $c^*_{opt} = 2.565$; $\lambda_{opt} = 0.4002$. Finally, for MV-optimality, we need to minimize $f(s, c^*)$ which corresponds to the larger of the two variance expressions. Routine calculations yield: $c^*_{opt} = 1 + \sqrt{2} = 2.4142$ and $s_{opt} = [1 - e^{-c^*} + c^{*2}e^{-c^*}]^{-1} = 0.6984$ and, hence, $\lambda_{opt} = 0.3016$.

## Appendix

**Proof of** (7.6.20). Note first the following chain of equivalent inequalities:

$$e^{-c^*} < Pe^{-a^*} + Qe^{-b^*} \tag{A.1}$$

$$\Longleftrightarrow c^* > -\log(Pe^{-a^*} + Qe^{-b^*}) \tag{A.2}$$

$$\Longleftrightarrow \phi(c^*) > \phi(-\log(Pe^{-a^*} + Qe^{-b^*})) \tag{A.3}$$

$$\Longleftrightarrow w\phi(a^*) + (1-w)\phi(b^*) > \frac{e^{-\log(Pe^{-a^*}+Qe^{-b^*})} - 1}{-\log(Pe^{-a^*} + Qe^{-b^*})} \tag{A.4}$$

$$\Longleftrightarrow \frac{P(1-e^{-a^*}) + Q(1-e^{-b^*})}{Pa^*e^{-a^*} + Qb^*e^{-b^*}} > \frac{(1 - Pe^{-a^*} - Qe^{-b^*})}{[(Pe^{-a^*} + Qe^{-b^*})(-\log(Pe^{-a^*} + Qe^{-b^*}))]} \tag{A.5}$$

$$\Longleftrightarrow [(Pe^{-a^*} + Qe^{-b^*})(-\log(Pe^{-a^*} + Qe^{-b^*}))] > Pa^*e^{-a^*} + Qb^*e^{-b^*} \tag{A.6}$$

$$\Longleftrightarrow [1 + \frac{P}{Q}e^{b^*-a^*}][b^* - \log Q - \log(1 + \frac{P}{Q}e^{b^*-a^*})] > b^* + \frac{Pa^*}{Q}e^{b^*-a^*} \tag{A.7}$$

$$\Longleftrightarrow (Q + Pe^{b^*-a^*})\log(Q + Pe^{b^*-a^*}) < P(b^* - a^*)e^{b^*-a^*} \tag{A.8}$$

$$\Longleftrightarrow (Q + Pe^{x^*})\log(Q + Pe^{x^*}) < Px^*e^{x^*};\ x^* = b^* - a^* > 0. \tag{A.9}$$

Define $\delta(x^*) = Px^*e^{x^*} - (Q + Pe^{x^*})\log(Q + Pe^{x^*})$, $x^* > 0$. It is easy to verify that $\delta(x^*)$ is increasing in $x^*$ since $x^* = \beta^* x$ and $\beta^*$ are positive. Further, $\delta(0) = 0$. Hence, the result follows.

# References

Abdelbasit, K. M. and Plackett, R. L. (1983). Experimental design for binary data. *Journal of the American Statistical Association* **78**, 90–98.

Agrawal, H. L. (1966). Some systematic methods of construction of designs for two-way elimination of heterogeneity. *Calcutta Statistical Association Bulletin* **15**, 93–108.

Atkinson, A. C. and Cook, R. D. (1995). D-optimum designs for heteroscedastic linear models. *Journal of the American Statistical Association* **90**, 204–212.

Bagchi, S. and Shah, K. R. (1989). On the optimality of a class of row-column designs. *Journal of Statistical Planning and Inference* **23**, 397–402.

Bagchi, S., Mukhopadhyay, A. C. and Sinha, Bikas K. (1990). A search for optimal nested row-column designs. *Sankhya, Series B* **52**, 93–104.

Bhaumik, D. K. (1995). Optimality in the competing effects model. *Sankhya* **57**, 48–56.

Das, K., Mandal, N. K. and Sinha, Bikas K. (2000). de la Garza phenomenon re-visited: Part II: Optimal designs under heteroscedastic errors in linear regression. Submitted to *Statistics and Probability Letters.*

Ford, I., Torsney, B. and Wu, C. F. J. (1992). The use of a canonical form in the construction of locally optimal designs for non-linear problems. *Journal of the Royal Statistical Society,Series* **54**, 569–583.

Heiligers, B. and Sinha, Bikas K. (1995). Optimality aspects of Agrawal designs - Part II. *Statistica Sinica* **5**, 599–604.

Hedayat, A. S. and Raghavarao, D. (1975). 3-way BIB designs. *Journal of Combinatorial Theory, Series A* **18**, 207–209.

Hedayat, A. S., Yan, B. and Pezzuto, J. M. (1997). Modeling and identifying optimum designs for fitting dose-response curves based on raw optical density data. *Journal of the American Statistical Association* **92**, 1132–1140.

Khan, M. K. and Yazdi, A. A. (1988). On D-optimal designs for binary data. *Journal of Statistical Planning and Inference* **18**, 83–91.

Kiefer, J. C. (1975). Construction and optimality of generalized Youden designs. In *A Survey of Statistical Design and Linear Models.* Ed. J. N. Srivastava. North - Holland, Amsterdam. 333–353.

Mandal, N. K., Shah, K. R. and Sinha, Bikas K. (2000). de la Garze phenomenon re-visited. Submitted to *Metrika.*

Mathew, T. and Sinha, Bikas K. (2001). Optimal designs for binary data under logistic regression. *Journal of Statistical Planning and Inference* **93**, 295–307

Minkin, S. (1987). Optimal design for binary data. *Journal of the American Statistical Association* **82**, 1098–1103.

Minkin, S. (1993). Experimental design for clonogenic assessment in chemotherapy. *Journal of American Statistical Association* **88**, 410–420.

Morgan, J. P. (1996). Nested Designs. In *Handbook of Statistics,* **13** (Design and Analysis of Experiments). Ed. S. Ghosh and C. R. Rao. North Holland, Amsterdam. 939–976.

Ozawa, K., Jimbo, M., Kageyama, S. and Mejza, S. (2001). Optimality and construction of incomplete split-block designs. To appear in *Journal of Statistical Planning and Inference.*

Pukelsheim, F. (1993). *Optimal design of experiments.* Wiley, New York.

Raghavarao, D., Federer, W. T. and Schwager, S. J. (1986). Characteristics for distinguishing balanced incomplete block designs with repeated blocks. *Journal of Statistical Planning and Inference* **13**, 151–163.

Raghavarao, D. and Zhou, B. (1998). Universal optimality of UE 3-designs for a competing effects model. *Communications in Statistics - Theory and Methods* **27**(1), 153–164.

Saharay, R. (1996). A class of optimal row-column designs with some empty cells. *Statistica Sinica* **6**, 989–996.

Sebastiani, P. and Settimmi, R. (1997). A note on D-optimal designs for a logistic regression model. *Journal of Statistical Planning and Inference* **59**, 359–368.

Shah, K. R. (2000). Optimal split-block designs. Unpublished Manuscript.

Shah, K. R. and Sinha, Bikas K. (1990). Optimality aspects of Agrawal designs. *Gujarat Statistical Review: Professor C. G. Khatri Memorial Volume* **7**, 214–222.

Shah, K. R. and Sinha, Bikas K. (1996). Row-column designs. In *Handbook of Statistics* **13** : Design and Analysis of Experiments. Ed. S.

Ghosh and C. R. Rao. North Holland, 903–938.

Shah, K. R. and Sinha, Bikas K. (2001a). Discrete optimal designs: Criteria and characterizations. To appear in *Recent Advances in Experimental Designs and Related Topics* (Proceedings of a Symposium held in Honour of Professor D. Raghavarao in Temple University in October, 1999). Nova Science Publishers, 115–133.

Shah, K. R. and Sinha, Bikas K. (2001b). Nested experimental designs. *Encyclopedia of Environmetrics.* Wiley, New York.

Singh, M. and Dey, A. (1979). Block designs with nested rows and columns. *Biometrika* **66**, 321–327.

Sitter, R. R. and Wu, C. F. J. (1993). Optimal designs for binary response experiments: Fieller-, D, and A criteria. *Scandinavian Journal of Statistics* **20**, 329–342.

Srivastava, J. N. (1978). Statistical design of agricultural experiments. *Journal of Indian Society of Agricultural Statististics* **30**, 1-10.

Wu, C. F. J. (1988). Optimal design for percentile estimation of a quantal response curve. In *Optimal Design and Analysis of Experiments*, Eds. Y. Dodge, V. Federov and H. P. Wynn. Elsevier, Amsterdam, 213–223.

Yeh, C. M. (1986). Condition for universal optimality of block designs. *Biometrika* **73**, 701–706.

# Author Index

# Subject Index

## Lecture Notes in Statistics

For information about Volumes 1 to 109, please contact Springer-Verlag

110: D. Bosq, Nonparametric Statistics for Stochastic Processes. xii, 169 pp., 1996.

111: Leon Willenborg and Ton de Waal, Statistical Disclosure Control in Practice. xiv, 152 pp., 1996.

112: Doug Fischer, Hans-J. Lenz (Editors), Learning from Data. xii, 450 pp., 1996.

113: Rainer Schwabe, Optimum Designs for Multi-Factor Models. viii, 124 pp., 1996.

114: C.C. Heyde, Yu. V. Prohorov, R. Pyke, and S. T. Rachev (Editors), Athens Conference on Applied Probability and Time Series Analysis, Volume I: Applied Probability In Honor of J.M. Gani. viii, 424 pp., 1996.

115: P.M. Robinson and M. Rosenblatt (Editors), Athens Conference on Applied Probability and Time Series Analysis, Volume II: Time Series Analysis In Memory of E.J. Hannan. viii, 448 pp., 1996.

116: Genshiro Kitagawa and Will Gersch, Smoothness Priors Analysis of Time Series. x, 261 pp., 1996.

117: Paul Glasserman, Karl Sigman, and David D. Yao (Editors), Stochastic Networks. xii, 298 pp., 1996.

118: Radford M. Neal, Bayesian Learning for Neural Networks. xv, 183 pp., 1996.

119: Masanao Aoki and Arthur M. Havenner, Applications of Computer Aided Time Series Modeling. ix, 329 pp., 1997.

120: Maia Berkane, Latent Variable Modeling and Applications to Causality. vi, 288 pp., 1997.

121: Constantine Gatsonis, James S. Hodges, Robert E. Kass, Robert McCulloch, Peter Rossi, and Nozer D. Singpurwalla (Editors), Case Studies in Bayesian Statistics, Volume III. xvi, 487 pp., 1997.

122: Timothy G. Gregoire, David R. Brillinger, Peter J. Diggle, Estelle Russek-Cohen, William G. Warren, and Russell D. Wolfinger (Editors), Modeling Longitudinal and Spatially Correlated Data. x, 402 pp., 1997.

123: D. Y. Lin and T. R. Fleming (Editors), Proceedings of the First Seattle Symposium in Biostatistics: Survival Analysis. xiii, 308 pp., 1997.

124: Christine H. Müller, Robust Planning and Analysis of Experiments. x, 234 pp., 1997.

125: Valerii V. Fedorov and Peter Hackl, Model-Oriented Design of Experiments. viii, 117 pp., 1997.

126: Geert Verbeke and Geert Molenberghs, Linear Mixed Models in Practice: A SAS-Oriented Approach. xiii, 306 pp., 1997.

127: Harald Niederreiter, Peter Hellekalek, Gerhard Larcher, and Peter Zinterhof (Editors), Monte Carlo and Quasi-Monte Carlo Methods 1996. xii, 448 pp., 1997.

128: L. Accardi and C.C. Heyde (Editors), Probability Towards 2000. x, 356 pp., 1998.

129: Wolfgang Härdle, Gerard Kerkyacharian, Dominique Picard, and Alexander Tsybakov, Wavelets, Approximation, and Statistical Applications. xvi, 265 pp., 1998.

130: Bo-Cheng Wei, Exponential Family Nonlinear Models. ix, 240 pp., 1998.

131: Joel L. Horowitz, Semiparametric Methods in Econometrics. ix, 204 pp., 1998.

132: Douglas Nychka, Walter W. Piegorsch, and Lawrence H. Cox (Editors), Case Studies in Environmental Statistics. viii, 200 pp., 1998.

133: Dipak Dey, Peter Müller, and Debajyoti Sinha (Editors), Practical Nonparametric and Semiparametric Bayesian Statistics. xv, 408 pp., 1998.

134: Yu. A. Kutoyants, Statistical Inference For Spatial Poisson Processes. vii, 284 pp., 1998.

135: Christian P. Robert, Discretization and MCMC Convergence Assessment. x, 192 pp., 1998.

136: Gregory C. Reinsel, Raja P. Velu, Multivariate Reduced-Rank Regression. xiii, 272 pp., 1998.

137: V. Seshadri, The Inverse Gaussian Distribution: Statistical Theory and Applications. xii, 360 pp., 1998.

138: Peter Hellekalek and Gerhard Larcher (Editors), Random and Quasi-Random Point Sets. xi, 352 pp., 1998.

139: Roger B. Nelsen, An Introduction to Copulas. xi, 232 pp., 1999.

140: Constantine Gatsonis, Robert E. Kass, Bradley Carlin, Alicia Carriquiry, Andrew Gelman, Isabella Verdinelli, and Mike West (Editors), Case Studies in Bayesian Statistics, Volume IV. xvi, 456 pp., 1999.

141: Peter Müller and Brani Vidakovic (Editors), Bayesian Inference in Wavelet Based Models. xiii, 394 pp., 1999.

142: György Terdik, Bilinear Stochastic Models and Related Problems of Nonlinear Time Series Analysis: A Frequency Domain Approach. xi, 258 pp., 1999.

143: Russell Barton, Graphical Methods for the Design of Experiments. x, 208 pp., 1999.

144: L. Mark Berliner, Douglas Nychka, and Timothy Hoar (Editors), Case Studies in Statistics and the Atmospheric Sciences. x, 208 pp., 2000.

145: James H. Matis and Thomas R. Kiffe, Stochastic Population Models. viii, 220 pp., 2000.

146: Wim Schoutens, Stochastic Processes and Orthogonal Polynomials. xiv, 163 pp., 2000.

147: Jürgen Franke, Wolfgang Härdle, and Gerhard Stahl, Measuring Risk in Complex Stochastic Systems. xvi, 272 pp., 2000.

148: S.E. Ahmed and Nancy Reid, Empirical Bayes and Likelihood Inference. x, 200 pp., 2000.

149: D. Bosq, Linear Processes in Function Spaces: Theory and Applications. xv, 296 pp., 2000.

150: Tadeusz Caliński and Sanpei Kageyama, Block Designs: A Randomization Approach, Volume I: Analysis. ix, 313 pp., 2000.

151: Håkan Andersson and Tom Britton, Stochastic Epidemic Models and Their Statistical Analysis. ix, 152 pp., 2000.

152: David Ríos Insua and Fabrizio Ruggeri, Robust Bayesian Analysis. xiii, 435 pp., 2000.

153: Parimal Mukhopadhyay, Topics in Survey Sampling. x, 303 pp., 2000.

154: Regina Kaiser and Agustín Maravall, Measuring Business Cycles in Economic Time Series. vi, 190 pp., 2000.

155: Leon Willenborg and Ton de Waal, Elements of Statistical Disclosure Control. xvii, 289 pp., 2000.

156: Gordon Willmot and X. Sheldon Lin, Lundberg Approximations for Compound Distributions with Insurance Applications. xi, 272 pp., 2000.

157: Anne Boomsma, Marijtje A.J. van Duijn, and Tom A.B. Snijders (Editors), Essays on Item Response Theory. xv, 448 pp., 2000.

158: Dominique Ladiray and Benoît Quenneville, Seasonal Adjustment with the X-11 Method. xxii, 220 pp., 2001.

159: Marc Moore (Editor), Spatial Statistics: Methodological Aspects and Some Applications. xvi, 282 pp., 2001.

160: Tomasz Rychlik, Projecting Statistical Functionals. viii, 184 pp., 2001.

161: Maarten Jansen, Noise Reduction by Wavelet Thresholding. xxii, 224 pp., 2001.

162: Constantine Gatsonis, Bradley Carlin, Alicia Carriquiry, Andrew Gelman, Robert E. Kass Isabella Verdinelli, and Mike West (Editors), Case Studies in Bayesian Statistics, Volume V, xiv, 448 pp., 2001.

163: Erkki P. Liski, Nripes K. Mandal, Kirti R. Shah, and Bikas K. Sinha, Topics in Optimal Design, xii, 172 pp., 2002.

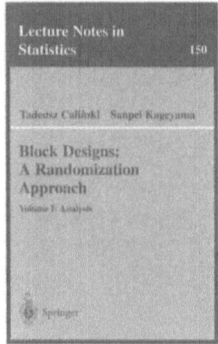

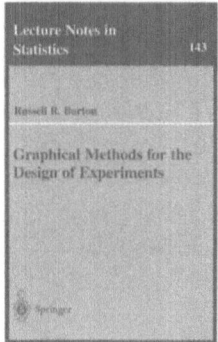

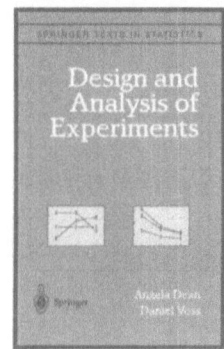